CW01207997

ORGANISATIONAL BEHAVIOUR IN HEALTH CARE

Organisational Behaviour in Health Care

The Research Agenda

Edited by
Annabelle L. Mark
and
Sue Dopson

Foreword by Rosemary Stewart

MACMILLAN Business

Selection, editorial matter and Chapters 1 and 17
© Annabelle L. Mark and Sue Dopson 1999

Individual chapters (in order) © Sandra Dawson; Peter Spurgeon; Edward Peck and Jenny Secker; Alexandra Harrison, Amy Pablo and Marja Verhoef; Lynn Ashburner and Katherine Birch; Virginia Morley, Nicki Spiegal, Faruk Majid and Priscilla Laurence; Lorna McKee, Gordon Marnoch and Nicola Dinnie; Bie Nio Ong and Rita Schepers; Beverly Alimo-Metcalfe; Graeme Currie; Annabelle L. Mark; Louise Fitzgerald, Ewan Ferlie, Martin Wood and Chris Hawkins; Steve Cropper; Frank Blackler, Andy Kennedy and Mike Reed; John Øvretveit 1999

All rights reserved. No reproduction, copy or transmission of this publication may be made without written permission.

No paragraph of this publication may be reproduced, copied or transmitted save with written permission or in accordance with the provisions of the Copyright, Designs and Patents Act 1988, or under the terms of any licence permitting limited copying issued by the Copyright Licensing Agency, 90 Tottenham Court Road, London W1P OLP.

Any person who does any unauthorised act in relation to this publication may be liable to criminal prosecution and civil claims for damages.

The authors have asserted their rights to be identified as the authors of this work in accordance with the Copyright, Designs and Patents Act 1988.

First published 1999 by
MACMILLAN PRESS LTD
Houndmills, Basingstoke, Hampshire RG21 6XS
and London
Companies and representatives throughout the world

ISBN 0-333-74555-8

A catalogue record for this book is available from the British Library.

This book is printed on paper suitable for recycling and made from fully managed and sustained forest sources.

10 9 8 7 6 5 4 3 2 1
08 07 06 05 04 03 02 01 00 99

Printed in Great Britain by
Antony Rowe Ltd
Chippenham
Willshire

For our daughters

Louise Mark
and
Sophie, Emily and Grace Dopson

Contents

Foreword by Rosemary Stewart ix

Acknowledgements xiii

Notes on the Contributors xv

1 Introduction 1
  Annabelle L. Mark and Sue Dopson

2 Managing, Organising and Performing in Health Care:
  what do we know and how can we learn? 7
  Sandra Dawson

3 Organisational Development: from a reactive to
  a proactive process 25
  Peter Spurgeon

4 Methodology and Marketing: making organisational
  behaviour research irresistible in health care 35
  Edward Peck and Jenny Secker

5 The Consumer's Role in Co-ordination: making sense
  of transitions in health care 47
  Alexandra Harrison, Amy Pablo and Marja Verhoef

6 Professional Control Issues between Medicine
  and Nursing in Primary Care 63
  Lynn Ashburner and Katherine Birch

7 Enabling Leaders to Change: interventions with
  established GP principals through a mid-career
  break scheme 77
  Virginia Morley, Nicki Spiegal, Faruk Majid and
  Priscilla Laurence

8 Medical Managers: puppetmasters or puppets?
  Sources of power and influence in clinical directorates 89
  Lorna McKee, Gordon Marnoch and Nicola Dinnie

Contents

9	Variations on a Theme: clinicians in management in England and the Netherlands *Bie Nio Ong and Rita Schepers*	117
10	Leadership in the NHS: what are the competencies and qualities needed and how can they be developed? *Beverly Alimo-Metcalfe*	135
11	The Influence of Middle Management upon Emergent Strategy: a case for more microempirical studies *Graeme Currie*	153
12	MAPS for PAMS: managerial and professional solutions for professions allied to medicine *Annabelle L. Mark*	169
13	Evidence into Practice? An exploratory analysis of the interpretation of evidence *Louise Fitzgerald, Ewan Ferlie, Martin Wood and Chris Hawkins*	189
14	Value Critical Analysis and Actor Network Theory: two perspectives on collaboration in the name of health *Steve Cropper*	207
15	Organising for Incompatible Priorities *Frank Blackler, Andy Kennedy and Mike Reed*	223
16	Evaluating Interventions to Health Organisation *John Øvretveit*	243
17	Conclusion *Annabelle L. Mark and Sue Dopson*	255
Index		263

Foreword

This book fulfils the purposes described in Chapter 1. It includes a wide variety of topics and makes many suggestions for future research. A question that the original symposium at Middlesex University in 1998 did not seek to address is: 'what are the criteria for saying that an area is a fruitful one for future research into organisational behaviour in health care?'. Nor were two related questions, 'how are areas for research identified?' and 'how is health care defined?' examined. This Foreword will not discuss the last question, which is included to remind readers, including future conference organisers, to consider whether an answer is needed for their purposes.

Drawing attention to these omissions is not a criticism of the aims of the symposium, but is aimed at encouraging readers to reflect on how the research agenda on organizational behaviour in health care has been, and is, interpreted. My own concerns about this were first aroused by the disappointing difficulties that I found in tackling an invitation to select about 25 articles on health care management; although this is a different, though related, field of research, I think the difficulties are also relevant to the subject of the present book. One reason why my selection proved to be so disappointingly difficult is that, unlike the situation in many other fields of social research, there is little sense of a developing body of knowledge in health care management, which starts from early research that is recognised as the foundation of later work, and from which major new developments can be traced. One practical indication of such a body of knowledge is the common recognition of the importance of particular works, and the inclusion of those that are in article form, as a collection published in book form: a second indication is how far back citations go. Another, and related, reason for the difficulty was that of finding articles that met the criteria of continuing relevance for an international readership.

Why is there no body of knowledge that is used as a foundation for current research? Does it matter, except for the task that I had of compiling a reader? One possible answer to both questions is that the research is more applied and more tied to its national characteristics than research in industry or public administration. This could help to explain why this was predominantly a UK symposium at Middlesex, despite the aim of making it an international one. Another answer, and one which would probably be favoured by some of the contributors, is that they draw upon the foundations of organisational behaviour to provide the theoretical underpinnings of their research and that is sufficient. A further answer is that both health care management and organisational behaviour in health care cover very wide areas, so that one cannot expect to find a body of knowledge that is relevant, but rather many different bodies of knowledge within different sub-sections, such as the sociology of professions. If the latter is true, it raises questions about the rationale for this symposium that included so many different subjects: a pragmatic rationale may be that there is not much research going on into organisational behaviour in health care, so it is unrealistic for a symposium to restrict the subjects within it.

There are, then, reasonable explanations for the lack of a well grounded area of research within organisational behaviour in healthcare; but I find the question, 'does it matter?', more difficult to answer and hence more worrying. It matters in that it makes it more difficult to identify the criteria for saying that an area is a fruitful one for research. So how do we tell that it is? My personal answer, developed over many years of qualitative research, comprises a number of scientific considerations, followed by the necessary pragmatic ones. Does the subject interest me? Does it develop from my previous work? Is the research likely to advance knowledge in this area? (In my view some, perhaps even much, research is done that produces findings that are trivial and so do not advance knowledge in any worthwhile way.) What previous work can it build on? The more pragmatic criteria, in my personal order of importance, are as follows:

- Is the research feasible? In my view some of the more interesting and important areas in social research can be too difficult to tackle successfully.
- Will the findings be of practical value? Here I distinguish between more applied and more theoretical areas and for the former I think it is important to ask whether this is an area where the research is likely to be used or one where what is needed is not more knowledge but political will, using 'political' in the broadest, non-party sense.
- Can adequate access be obtained for the field work?
- Can funding be found and, if so, are the conditions of the funding acceptable? This applies especially to restrictions on publication.

- Will the relevance of the findings date quickly and, if so, is the research likely to be complete and written up in time?

Another possible pragmatic consideration for young researchers trying to build their reputation is whether it is a fashionable area of research or one that they believe they could make fashionable.

It is hoped that some of the younger readers will find these personal criteria helpful in thinking about fruitful areas for research in organisational behaviour in health care. Perhaps my anxieties about the state of research in health care management and in organisational behaviour in health care arise from unrealistically scholarly ideas? It may be that such research is necessarily pragmatic and responsive to the interests of research funders and to the many factors, particularly political ones, that are changing the perception and practice of health care. Perhaps most of the research is, and should be, research consultancy, as a justification for taking up busy people's time. But I still feel concerned about what I think is a lack of intellectual development within the field of research into organisational behaviour in health care and in health care management.

ROSEMARY STEWART
Templeton College

Acknowledgements

The following people served on the original conference advisory panel and were involved in the refereeing process for the contributions now appearing in this book: Alan Cowling (Middlesex University), David Sims (Brunel University), Louise Fitzgerald (Warwick University Business School), Tony White (Bournemouth University), Lyn Ashburner (Keele University), Kevin Smith (McMaster University, Canada) and John Øvretveit (Nordic School of Public Health, Sweden). Thanks also to Charlotte Binden and Mike Chapman for Secretarial and IT support respectively.

Notes on the Contributors

Beverly Alimo-Metcalfe is Professor of Leadership Studies at the Nuffield Institute of Health, Leeds University, and a Chartered Occupational/Industrial Psychologist and a Fellow of the British Psychological Society. She has worked extensively in public and private sector organisations in the assessment and development of leadership. She also has a strong interest in gender in relation to this topic. Her work for the Local Government Management Board on the nature of transformational leadership in the UK public sector has been extended to the NHS, producing a new model of transformational leadership.

Lynn Ashburner lectures in management at the Centre for Health Planning and Management at Keele University, and was previously at Nottingham University and Warwick Business School. She has carried out research in both public and private sectors and worked in organisational development at Pilkington. Current interests include the understanding of large-scale organisational change, relating to changes in health care services, and the changing role of primary care. She has published widely and is co-author of *The New Public Management in Action*.

Katherine Birch is a Research Fellow at the Centre for Health Planning and Management at Keele University. She worked in NHS management before moving into research and teaching and is currently working on a range of projects concerned with the development of primary care services.

Frank Blackler is Professor of Organisational Behaviour at the University of Lancaster. His research interests have included organisational change, work systems design, new technologies and the management of innovation. Most recently he has pioneered the application of activity theory to organisation studies, utilising the approach to feature central processes in collaboration, knowledge management and organisational learning.

Steve Cropper is a Senior Lecturer in Management at the Centre for Health Planning and Management at Keele University. He has worked at both Sussex and Strathclyde Universities developing problem structuring, decision support and group decision support methods. He is co-editor with Paul Forts of *Enhancing Health Services Management: The Role of Decision Support Systems.*

Graeme Currie is a Lecturer in Health Services Management at the University of Nottingham Business School. His main research interest lies with the role of middle managers in strategic change. Research projects focus upon the NHS, for which he has obtained funding, include the impact of IT upon the role of middle managers and the role of the middle manager in the implementation of human resource strategy.

Sandra Dawson is KPMG Professor of Management Studies and Director of the Judge Institute of Management Studies, and a Fellow of Jesus College, at the University of Cambridge. She has been both a chairperson and member of health authorities and serves on the Senior Salaries Review Body. She acts as a consultant to a wide range of organisations in the public and private sector. She has written numerous articles as well as *Analysing Organisations, Safety at Work: The Limits of Self-regulation* and *Managing in the NHS: A Study of Senior Executives.*

Nicola Dinnie is a researcher within the Department of Management Studies at the University of Aberdeen. She has previously researched and published with her colleagues Lorna MacKee and Gordon Marnoch on the roles of clinical directors in Scotland, quality of life issues, and biomedical innovation networks. She is currently establishing a network of biomedical ethics researchers.

Sue Dopson is a Lecturer in Management Studies and Fellow in Organisational Behaviour at Templeton College, Oxford. After a brief career in personnel in the NHS she became an academic and has researched many aspects of health care and organisational behaviour, in particular changes in the management of the NHS and the public sector.

Ewan Ferlie is Professor of Public Services Management at Imperial College Management School, University of London. He previously worked at the Centre for Corporate Strategy and Change, Warwick Business School, where he was Deputy Director. His research interests centre on service delivery and organisation issues in health care, with a focus on innovation, organisational change and the behaviour of professionals. He is currently working on the implementation of evidence-based medicine ideas in clinical settings.

Louise Fitzgerald worked in a variety of personnel management posts before becoming a university lecturer. In 1986 she moved to Warwick University as a Senior Lecturer; and in 1998 moved to her present chair at City University. Her research and publications centre on the management of change in the health care sector.

Alexandra Harrison is a health care professional, manager, researcher and teacher. She is the Director of Graduate Clinical Education for the Faculty of Medicine at the University of Calgary in Alberta, Canada. Her teaching and research interests include the design, management and evaluation of health service organisations.

Chris Hawkins is a Research Fellow at the Centre for Creativity, Strategy and Change, Warwick Business School. She has a background of working as a physiotherapist and physiotherapy manager in the NHS until 1994. She has an MA in strategic decision-making and works in the area of clinical behaviour change in primary care.

Andy Kennedy is a Fellow at the King's Fund, London, an independent foundation concerned to improve the health of citizens. Andy is director of two of the King's Fund leadership development programmes for senior managers and clinician managers, designed to address such issues as the meaning of collaboration, professional autonomy, social justice and the relationship of health care organisation to healthy lives.

Priscilla Laurence is a consultant working in health care, specialising in personal and organisational development. She joined the NHS in 1991 working in a large London teaching hospital. Prior to this she worked as a management consultant in strategic development within the public and voluntary sectors. She has subsequently worked at the South Thames Regional Office of the NHS Executive until 1996 and is now working independently.

Faruk Majid has been a general practitioner in Lewisham since 1989. He is now part of a four-partner fundholding practice with a patient population of 10 000. His main interests are GP education and quality standards in practice, commissioning issues and service development. He has been Lecturer in the Department of General Practice at King's College, London.

Annabelle L. Mark is a Senior Lecturer in Health Management and Organisational Behaviour at Middlesex University Business School, a Research Associate of Oxford Health Care Management Institute at Templeton College, and a Fellow of the Institute of Health Services Management. Following a ten-year career as a manager in the NHS she now undertakes consultancy and research and writes almost exclusively on health management issues. Her current interests are changing roles for the professions in health care, organisational discourse and the management of emotion, and the management of demand for health care. (*She has also published as Annabelle Mark but is not to be confused with Annabelle May, who also writes about healthcare.*)

Gordon Marnoch is Director of Postgraduate Programmes, Department of Management Studies, University of Aberdeen. His previous research focuses on the impact of general management in the NHS, roles and responsibilities of clinical directors, management and innovation in primary care, corporate governance, and devolved management, in NHS Trusts. His published work includes joint authorship of *Just Managing: power and culture in the NHS* and *Doctors and Management in the NHS*.

Lorna McKee is Director of Research, Department of Management Studies, University of Aberdeen. She is also Co-Director of the leadership programme in primary care (with G. Marnoch) funded by the Scottish Office Home and Health Department. Her research interests include health care management, work/family balance issues and bioethics. She is co-author of *Shaping Strategic Change: Managing Change in the NHS* (with E. Ferlie and A. Pettigrew), and has worked as a management consultant in the NHS over a number of years.

Virginia Morley is a Senior Lecturer in Primary Care Development in the Department of General Practice and Primary Care at Guy's, King's and St Thomas' and she is also an independent health care consultant. She has a particular interest in the organisational development of primary health care in the UK.

Bie Nio Ong is Professor of Health Services Research at the Centre for Health Planning and Management at Keele University. Her research interests include the development of medical management and mobility of medical labour between countries. She was a non-executive director of an acute hospital trust and has recently been appointed to a Health Authority board.

John Øvretveit is Professor of Health Policy and Management at the Nordic School of Public Health in Gothenburg, Sweden, and director of the postgraduate diploma in healthcare quality at Bergen University. He has worked as a clinician in the NHS and then as Director of the Health Services Centre at Brunel University. He publishes widely on healthcare quality, evaluation, interprofessional working and health reform.

Amy Pablo is Associate Professor and Director of the MBA in enterprise development in the Faculty of Management at the University of Calgary. She received her doctorate from the University of Texas at Austin and her current research centres on managerial decision-making and organisational outcomes, with a focus on the influence of risk on such processes and an examination of strategic alliances as a specific decision domain.

Edward Peck is Director of the Centre for Mental Health Services Development at King's College, London. Between 1984 and 1991 he was a manager of mental health services in the NHS. He undertakes research and consultancy work on mental health and the relationships between primary and secondary care. He has published numerous articles as well as *NHS Trusts in Practice* and is a contributor to the King's Fund Commission work on mental health services in London.

Mike Reed is Professor of Organisation Theory in the Department of Behaviour in Organisations at Lancaster University. His research interests include theoretical development in organisational analysis, changes to the expert division of labour and their implications for organisational forms, and the emergence of disorganised organisations in postmodernity. He is joint editor with Gibson Burrell of the journal *Organisation*.

Rita Schepers is Associate Professor in Social Medical Sciences in the Department of Health Policy and Management at Erasmus University,

Rotterdam. Her research focuses on the development of the division of labour in health care, on the developments in the interface between management and medicine, and between medicine and financiers, in particular in the Netherlands and Belgium.

Jenny Secker is a Senior Research Fellow with the Centre for Mental Health Services Development at King's College, London. She is a qualified psychiatric nurse and social worker and was awarded her PhD in 1991 for her study of social work training. Her research interests include the implementation of mental health policy and the application of qualitative methods in health services research.

Nicki Spiegal originally trained as a nurse but now works as a research and development facilitator, assisting multidisciplinary teams to develop their services in areas such as managing change, teambuilding, strategic planning and the development of learning and quality. She has a number of publications in this area, and is vice-chairperson of the Royal College of General Practitioners Commission on Primary Care.

Peter Spurgeon is Director of Research at the Health Services Management Centre at the University of Birmingham, where he was formerly Director. Initially a psychologist he also taught at London and Aston Universities and has worked in a range of sectors. He has interests in management learning, organisational development and decision-making, and is currently exploring the link between organisational culture and clinical risk management as well as identifying facilitators and inhibitors of change in clinical practice.

Marja Verhoef received her MSc from the State University of Utrecht and her PhD in epidemiology from the University of Calgary. She is an Associate Professor in the Department of Community Health Sciences at the University of Calgary. Her areas of expertise include research methods and psycho-social aspects of health and health care.

Martin Wood is Research Fellow at the Centre for Creativity, Strategy and Change in the Warwick Business School. His research interests currently include studying the heterogeneous processes, strategies and technologies of organising change within the NHS, particularly the translation of knowledge in relation to new organisational forms and practices.

1 Introduction

Annabelle L. Mark and Sue Dopson

The idea for this book arose from an international research symposium of the same title which took place at Middlesex Business School in January 1998. The symposium was created so that academics specialising in this field could meet and discuss their work in specific terms as well as debating broader yet related issues such as the use of methodologies and research design, and the problems and politics surrounding research efforts in health care. Rarely are such opportunities available within many of the mainstream management conferences that this group may contribute to, and one outcome is that further such opportunities are now planned for the future.

Another reason for the conference was to think about the fruitful future areas of research. This seemed especially timely given that the UK government is investing in a significant research agenda to determine the future provision of services. The emerging agenda recognises the potential contribution of organisational behaviour (OB) and organisational analysis (OA), but there do still appear to be pools of uncertainty as to the nature of these contributions. This book seeks to offer a flavour of research activities in this area in an attempt to clarify, not only what type of work is being done, but also what work might be undertaken.

Before seeking to briefly outline each of the papers selected from the conference, it is perhaps helpful to sketch the parameters of OB as a discipline within management studies. The boundaries around the discipline, like those of all developing disciplines, are constantly adjusting to growth in knowledge and understanding of the dynamic range of issues being studied. Nonetheless a very simple distinction can be made between OB, with its focus on the characteristics and processes of individuals and groups, and OA, where the focus is more on the organisation as a whole and characteristics such as structure, effectiveness and goals, as well as change and communication, which are also dealt with. Organisational development (OD) is in a sense the managerial application of information gained from the work coming out in OB and OA.

The book will appeal to those looking for an understanding of the discipline of OB and OA as applied to health services. Readers with a knowledge of health care will gain an understanding of the contribution that can be made and applied. It is also hoped that those looking for some new perspectives on future research may also gain some insight into the opportunities now emerging.

The book includes about half of the original conference contributions. The contributions from the keynote speakers are followed by papers looking at a series of current issues. The papers are set out to take the reader on the journey that the patient frequently makes through the health system, encountering different levels of complexity, processes and professional groupings. The two keynote papers, from Sandra Dawson and Peter Spurgeon, draw out many of the themes explored by later chapters. Sandra Dawson identifies a number of present-day challenges, notably the need to appreciate different worlds even within professional groups and the need to find ways of achieving accommodation and reducing fragmentation in the health sector, she also reflects on ways of relating organisational behaviour research to practice. Peter Spurgeon then looks at the challenges faced by the practical implementation of research through the medium of organisational development. He suggests that, as a strategy for organisations to respond to change, the focus on strategic management has neglected the contribution OD can make to operational management, for example through a focus on organisational culture. A better understanding of this, he suggests, could facilitate the management of innovation and risk, evidence-based medicine/management and the R&D agenda in general.

Edward Peck and Jenny Secker touch on a topic common to all the contributions: how to decide on a research design and a dissemination strategy. Chapter 4 reviews some of the methodological issues which continue to confound research, in particular the mismatch between research and managerial timescales; demands on the time of managers for qualitative data collection; the training of researchers in research methods; and the difficulty in conveying the 'thick descriptions' of qualitative data in a way which is valid but also accessible to both research and managerial communities.

Alexandra Harrison, Amy Pablo and Marja Verhoef offer a contribution from Canada on the role of consumers as co-ordinators of their own care. They report findings from a study that examined consumers' views about co-ordination and their experience of transition from one sector of health care to another. It is argued that this is an under-researched area crucial to the renegotiated role of health services in the future as an activity shared between patient and professional. It concludes that, in making sense of the transitions in health care, involving consumers should become a key component of organisational strategy.

Lynn Ashburner and Katherine Birch consider the problems of 'contested boundaries' in primary care where the development of new roles – notably that of the nurse practitioner – challenges the status quo. Contesting boundaries will of course be a feature of government changes to the NHS as set out in the Secretary of State's *The New NHS – Modern and Dependable*, published in 1997. In Chapter 6 they consider competing interpretations of the nurse practitioner role in the UK compared to that of their North American counterparts. They argue that such competing interpretations confound the development of primary care, because nurses see it as an extended role, government sees it as a cheap alternative to GPs, purchasers/commissioners see it as cost-effective and appropriate and GPs see it as an opportunity to share roles yet keep control over this enhancement to nurse activity.

The chapter by Virginia Morley and colleagues looks at stress at work for GPs and it considers developments for GPs in the provision of mid-career breaks. A feature of this chapter is a discussion of how the medical socialisation process and peer pressure prevent this group taking the option of a mid-career break. Given the developing role of general practice contained in the latest reforms of the NHS, understanding stress at work for GPs and evaluating strategies for helping is yet another important challenge, especially in the light of current recruitment problems.

The chapter by Lorna McKee and colleagues looks at the role of doctors in management. It is unusual in offering empirical insights into the Scottish experience and, interestingly, confirms much of the previous research outcomes identified from the wider UK experience. Bie Nio Ong and Rita Schepers also consider this theme in Chapter 9 and offer an international comparison. Both chapters reflect on how the involvement of doctors has changed and on the diversity of models involving clinicians in management, and debate the sustainability of these models in the ever-changing and contested environment of health care.

Beverly Alimo-Metcalfe reviews research on the nature of leadership within the NHS as part of a wider study of transactional and transformational leadership. The chapter looks specifically at the way leadership is perceived, through the differing lenses available via 360 degree feedback systems now being adopted in the NHS. Some significant findings emerge, not least some surprising gender differences which may have important consequences for future recruitment and selection.

Graeme Currie offers a study of middle managers in the development of strategic change. This is a rare empirical contribution to the debate on middle management which usually takes place at the level of speculation. This study found that the realisation of top management intent is an outcome of a process of political struggle and negotiation between the clinical directorate and middle managers. The author offers a range of methodological advice to those wishing to go down this neglected research track.

Annabelle L. Mark reviews an under-researched area of health care by considering the present and future for professions allied to medicine. Changes to their roles, activities and national and international environments are explored. The chapter concludes that a greater degree of flexibility will be required for the future organisation of these groups, and their relationships with others, if they are to deliver innovative professional care for patients.

Louise Fitzgerald and colleagues offer an insightful piece on the nature of evidence within the evidence-based medicine movement. Taking more general organisational models as a starting point, the chapter concludes that EBM assumes that knowledge itself is unproblematic and follows a linear process to implementation through passive recipients – this, the authors demonstrate, is not so. The team's empirical work suggests a number of interesting areas for research: for example, why do professionals count some evidence and discount other evidence; what are the effective ways to facilitate the translation of evidence between researchers and practitioners and across professional boundaries?

Steve Cropper uses a number of network theories to explore and analyse the complex web of relationships that we find in health care. He alerts us to the need to conceive of relationships beyond the level of individuals. He argues that building patterns of interdependencies and mutual interests will be critical for achieving a more collaborative environment.

In Chapter 15, Frank Blackler and his colleagues illuminate the contribution of post-modern perspectives to health care. In particular, they highlight the importance of recognising that most groups can only ever have a partial understanding of context, which is mediated by the vocabularies that they share. Without these insights, the authors argue, achieving real collaborative working will be confounded by a resort to simplicity.

In the final contribution, John Ovretveit looks at the problems associated with evaluation within this complex environment; in particular, he notes a number of intervening factors which make such activities problematic. The book concludes with a chapter by the editors which endeavours to draw out some of the themes and future areas for development in this particular research field.

2 Managing, Organising and Performing in Health Care: what do we know and how can we learn?

Sandra Dawson

This chapter reviews aspects of organisation, management and performance in health care with particular reference to the NHS in England. It addresses issues of knowledge and learning in respect of communities of academics and practitioners who are either health managers or medical doctors. The question which forms the subtitle of the chapter, 'what do we know and how can we learn?', will be applied to researchers, health managers and doctors.[1]

The discussion is in four parts. The first concerns approaches to management and organisation research; the second questions whether there is anything special about organisation and management in health systems as compared to others; the third illustrates the way research can relate to practice in health organisation and management; and the conclusion considers some lessons to be drawn from the chapter for researchers and practitioners.

This chapter builds on the findings of research undertaken with colleagues at Imperial College[2] (Dawson *et al.*, 1995) and Cambridge and Oxford Universities[3] (Dawson *et al.*, 1998) into management activities, career experience and knowledge generation and transfer in the NHS, and on personal experiences and insights gained while chairman of Riverside Mental Health NHS Trust between 1992 and 1995.

APPROACHES TO MANAGEMENT AND ORGANISATION RESEARCH

The study of management and organisation has been an enormous growth industry in the last 40 years. There has been an explosion in the number of studies and commentaries designed to increase knowledge about organisation and management. Many different approaches to generating knowledge and facilitating learning about management and organisation can be discerned, as Burrell and Morgan (1979) described. Faced with such an array of choice, researches learn that they need to make decisions about their position in relation to ontology, where they are guided by existing knowledge, ideology and learning, and to epistemology, as revealed in their research designs.

In order to locate the work on which this chapter is based within the broad fields of organisation and management research, it is described in terms of its position on two selected dimensions. The first, conceptual, reflects ontology and epistemology and relates to the orientation of the researcher towards the feasibility and desirability of generating knowledge which will be relevant to improving practice and enabling practitioners to learn. The second, contextual, relates to the nature of the subject matter, namely the extent to which any organisation under study is privately funded (predominantly through equity, debt and retained profits) or publicly funded (predominantly through taxation). These two dimensions have been put together to form the matrix shown in Figure 2.1.

The place of researchers on the horizontal axis (their orientation to the practical utility of their findings) is related to ideology. Three orientations

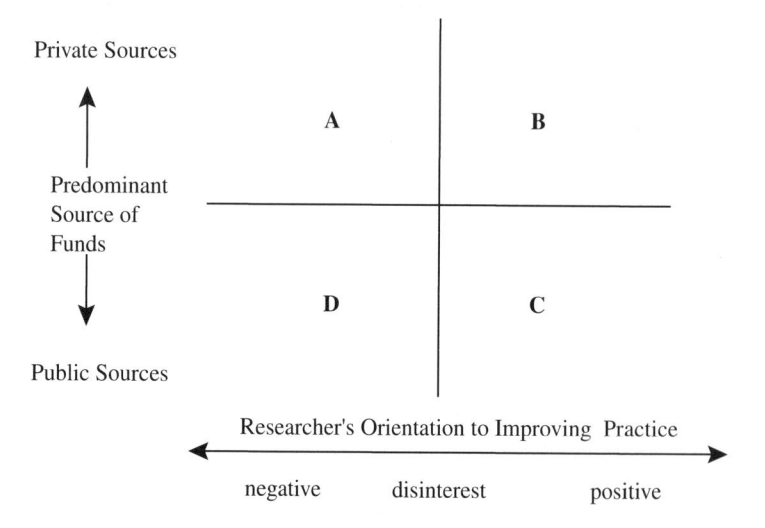

Figure 2.1 Approaches to the study of management and organisation

will be briefly described as three points on the continuous horizontal axis in Figure 2.1. As Burrell and Morgan illustrated in their description of radical organisation theory, some scholars conceptualise organisations as institutionalised systems of domination (Burrell and Morgan, 1979, pp. 365–92). In such cases, however organised or constituted, organisations are seen as arenas in which a minority benefit disproportionately and unjustifiably from their position and/or exercise disproportionate and unjustifiable control over others. Thus, if a researcher seeks to generate knowledge which will improve managerial or organisational performance, she or he is offering advice to those who are in established positions of power and, thereby, is acting to perpetuate unacceptable systems of inequality and domination. In laying bare the systems of inequality and domination, such researchers would be content to discover that their findings have been used to undermine the structure of domination.

Other scholars are sceptical of any particular orientation and prefer to reveal the nature of organisations and management without any prior commitment either to deny or to offer recommendations on how to enhance managerial learning or improve performance. Their orientation is to stop short of making recommendations for practitioners; they seek greater knowledge about management and organisation, in order to improve their own learning, although they are normally content if their work is found to be helpful to practitioners. Such a link between the knowledge they, as researchers, generate and the learning that others, as practitioners, enjoy is not, however, their primary concern.

A third approach, and the one adopted in this chapter, aims from the outset to design research projects which will lead to findings which are of practical use. The research cited has been funded by the Department of Health and the NHS Executive and the author has also been involved in the management of the NHS. Thus, as a practitioner, she has tried to generate knowledge from which she can learn more about the conceptual base and the practical application of her subject. The aim is to generate knowledge as a basis for learning how to improve performance, as well as to advance the subject conceptually and theoretically.

This does not mean that a researcher in this mode will adopt practitioners' prevailing and given views of the world. Indeed one of the benefits of independent research is that one can question assumptions and seek to introduce new ways of conceptualising and tackling problems and opportunities. The rigour of research design is all-important, regardless of one's ideological and practical orientations. Ensuring an independence of mind, that methods chosen for data collection and analysis are at an appropriate level of analysis (individual, group, department, division or organisation) and that the design will provide valid and realistic results are just as much cardinal requirements for research which aims to improve practice, as for research that falls elsewhere on the horizontal axis of Figure 2.1.

The second dimension, identified as the vertical axis in Figure 2.1, is the type of political economic context in which the organisations under study, operate. A distinction can be made between organisations which draw their funds predominantly from private sector investors, creditors and customers in a relatively free market for capital and revenue and those which derive their funds from national or local government or their agencies, which in turn will secure funds through taxation, albeit often supplemented by income derived from direct selling of goods and services to customers. While all organisations have some features in common, differences in political economy, which reflect core values about services as public goods and result in a separation of the customer from the payer, create differences in the environment in which managers perform. Once again we are not dealing with discrete categories, but with differences of degree. There are government regulations and subsidies, sponsorship and other forms of state involvement in private markets and there are private funds available in systems which are largely funded through the public purse. Taking the domain of health in England, this chapter will focus on organisations which are predominantly publicly funded.

It may be noticed that this contextual dimension has been described in terms of sources of funds rather than suppliers of service. Reforms in the NHS in the 1980s and 1990s have included many aspects of private markets: for example, compulsory competitive tendering, market testing, increased use of co-payments for goods and services, the private finance initiative to attract private capital, and a significant expansion of private and voluntary provision in community and social care, as well as the creation of quasi-autonomous NHS trusts. There is now a significant and diverse market in which priorities are determined and goods and services are purchased; but the purse from which core services are purchased is still predominantly held by public state agencies.

Accepting that the two dimensions in Figure 2.1 refer to differences of degree, the work which is the subject of this chapter falls within segment C. It draws on research which has been designed and undertaken in order to enhance knowledge and learning in the academic community about management and organisation, and to improve practice and performance in publicly funded health organisations, with a particular focus on the English NHS. Having located this work conceptually and contextually in the broad field of studies of management and organisation, there now follows a more specific discussion of the context in terms of a second general issue, the nature of health systems.

IS THERE ANYTHING SPECIAL ABOUT THE ORGANISATION AND MANAGEMENT OF HEALTH SYSTEMS?

Is health just another commodity, another collection of goods and services, perhaps more complex than transport or electrical goods, but fundamentally the same? In order to understand more about the way health systems may be managed or organised more effectively, do we need to study only health organisations, or no health organisations, or a few health organisations alongside others? Is there anything special about health organisations which limits the extent to which we can learn from other sectors?

Let us begin by stating the obvious: health systems are highly complex. It is all too easy to focus on small parts and to miss critical and unexpected linkages between them. Complexity reflects several key characteristics which are identified in Figure 2.2. In this figure the complexity is shown at a high level of generality in terms of the involvement of five different spheres which create the dynamics of supply, demand and political involvement in health. One could focus on any one of the five spheres and it would itself be revealed as a highly complex set of players, relationships, power, influence, change and uncertainty. For the purposes of this chapter, however, we will stay at the macro level and consider, first, supply.

Three spheres stand out for their significance for the supply side of health systems. Between them they provide an infinite supply of ideas, actions, technologies, procedures and drugs which may be applied to health. First, the scientific sphere. Endeavours to improve health are frequently grounded in scientific and technological knowledge from diverse fields. The results of research on genetics, biotechnology and pharmacology have come to be regarded as equally important as those from anatomy, physiology, medical engineering, psychology and device

12 Managing, Organising and Performing

industrial sphere	**scientific sphere**	**professional sphere**
powerful global industrial investment	scientific & technological knowledge & applications	occupational diversity professional dominance

infinite supply

complexity

infinite demand — political involvement

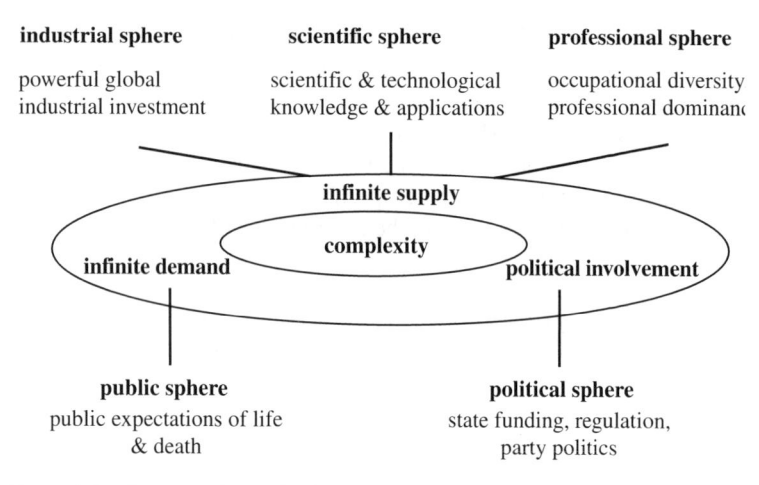

public sphere — public expectations of life & death

political sphere — state funding, regulation, party politics

Figure 2.2   Sources of complexities of health systems

design. Each field has its own structure of knowledge. Managing the interface between fields is a major challenge in which central players are universities, medical research charities and other funders, and large and powerful industrial sectors. All these parties work on a transnational scale.

The industrial sphere is important, not only for its support for research, but also for its global manufacturing, distribution and marketing activities. Manufacturers and distributors of drugs, medical technologies and consultancy advice are large and powerful players in global as well as national economies. The scale and growth of the health care industry is remarkable. It has been estimated that, at the beginning of the twentieth century, the health care industry represented US $20 billion in a US $1 trillion global economy. At the end of the century, the health care industry represents about US $3 trillion in a US $40 trillion economy. During the next century, if medical expenditure increases by 3 per cent per annum in real terms, global medicine will be close to a US $48 trillion industry in a US $500 trillion world economy.[4] The industry's importance as an attractor of capital investment, as an employer and as a purchaser of goods and services creates an environment in which it can be tremendously powerful in lobbying governments, and very important in influencing patterns of professional practice and individual consumption of health care.

Turning to the third sphere of supply identified in Figure 2.2, we have strong and determined professions and occupations who are well represented in England through associations such as the Royal Colleges and other professional groupings and trade unions. Securing health and fighting illness are labour-intensive activities which rely heavily on specialist professional occupations including diverse branches of medicine, nursing and therapy, as well as a wide array of managerial, administrative and clerical functions. The strength and diversity of professional groups and the complexity of the tasks which constitute health care mean there is a great tension between the required levels of expert specialisation and effective multiprofessional working.

The three spheres of industry, science and the professions only occasionally act in concert, but the combined effect of their frequently independent actions is to create an infinite supply of things, ideas and practices which could be realised and might improve health. This infinite supply, or the idea that supplies could be forthcoming if funding was available, flood onto a highly receptive market-place. The subject matter of health is life and death. Not surprisingly, therefore, health is seen to be of outstanding importance to patients and clients. Even more important, in terms of size of the market, is the fact that every citizen and their children, parents, friends and colleagues are 'potential', if not actual, patients and clients.

The fourth sphere, creating complexity is that of the involvement of the public. The manner in which public expectations of rights to health, quality of life and longevity are represented reflects different cultures, institutional structures and resource availability and allocation mechanisms. Nonetheless we can be fairly sure that, where the subject matter is life and death, we encounter what may be one of the few universal truths: that is, if given an alternative option, few people would vote for death and ill health for themselves or their family and friends. As fields of knowledge advance, so more is offered which promises to improve morbidity and quality of life, and lessen mortality rates. This frequently makes discussions about health and health care highly charged and inevitably political. Diverse and increasingly articulate pressure groups representing patients, clients and their families and communities join in debate involving health professionals, the press, local and national politicians and health service managers.

The fifth sphere in Figure 2.2 is that of politics and the state. Regardless of the detailed organisational arrangements for the provision of health care, government is always involved as a regulator and provider of 'last resort', if not as a primary purchaser or provider. Furthermore the life and death issues which touch us all make health a highly politically sensitive issue, even if the government is not a major funder. Party politicians know that health is an irresistible arena to proclaim victory and an unavoidable arena for vilification when fear of restriction of services or choice is generated, or promises cannot be kept, or incompetence or insensitivity is revealed. Consider, for example, the strong role played by political, as well as industrial, professional and labour interest groups, in opposing the Clinton health reforms in 1993 in the USA (Lamm, 1994; West *et al.*, 1996). The fiercely political nature of any discussion on health reform in the USA is discussed by Dionne and Hacker (1997) who argue that the 1993 efforts were but one small part of a historical drama which has been going on since President Roosevelt promised compulsory health insurance in his 1912 presidential campaign. And this is the story of a country where the state has a much less direct role in health care than in England.

In the UK, the state's role as a major funder adds considerably to political sensitivities. This was illustrated by the public discussion of the case of the 10-year-old girl with leukaemia, who became known as 'child B', when her father refused to accept the advice of doctors against further treatment and took his health authority to court for refusing to fund chemotherapy and a second bone transplant (Richmond, 1995; Sellman, 1997; Entwistle *et al.*, 1996). All senior managers in the NHS, even those employed in quasi-independent NHS trusts, know the great diversion and anxiety involved in being the subject of a Parliamentary Question or an interrogation in the Public Accounts Committee, let alone being summoned to court to account for decisions about priority setting and managerial process.

A further aspect of political complexity is that in any nation's health system there is always tension between a focus on securing the best value for money in terms of the immediate demands to improve a given individual's health, requiring investment in better diagnosis and more effective treatment of illness, and a focus on securing an improvement in the health of a given population. If resources were endless, tensions between the two would be more or less insignificant. As it is, with limited funding (whether state, corporate or private) difficult balances have to be determined between health promotion, disease prevention and a sickness service. This is reflected in discussions which take place throughout the world – whatever a country's GDP – about how much of a health budget can be applied to environmental investment to secure safer air, food and water, how much can be spent on securing more effective inter-agency working between housing, social services, education and health to promote healthy lifestyles, and how much on securing reductions in waiting lists for elective surgery or the purchase of the latest diagnostic equipment in hospitals.

Figure 2.2 and the accompanying discussion have identified five sources of complexity in health systems at a macro level. If we return to the question of whether management and organisation in health is a special case, we can find individual parallels for each source of complexity in other sectors. If we take just one sphere at a time, health is not so special. For example, we have high political involvement in education, critical industrial investment in major capital projects or global markets for consumer goods, a crucially important multidisciplinary science base for information technology, and an undoubted professional dependence and dominance in the law. The difference for managers and organisations in health is that they experience complexity on all these four fronts and in addition they hold the trump card (or maybe the joker in the pack) of complexity in the simple appeal of their subject matter to life and death.

To ask a number of seemingly simple questions of any health system – who funds, who provides, who benefits, who controls? – is to reveal answers which are never straightforward, always contestable and only resolvable (within the limits set by different politico socioeconomic cultural contexts) through an interplay between the five spheres of science, industry, professions, politicians and the public identified in Figure 2.2.

Thus, in terms of comparisons with other sectors, the challenges for management and organisation in health are similar to those for managers in other complex businesses. However, given the diversity and strength of the pressures on supply, demand and political involvement identified in Figure 2.2, the challenges to those involved in health are particularly great. Organisation and management in health should not always be studied as a special case. A great deal can be learnt from other sectors, but there is also considerable merit in focusing some studies specifically on health.

Three examples of empirical research which have sought to understand more about the particular challenges for management and organisation in health will now be considered, together with the recommendations which were made about the way the challenges could be met and the participating managers learn.

ILLUSTRATIONS OF THE CONTRIBUTION OF MANAGEMENT AND ORGANISATIONAL RESEARCH TO IMPROVING HEALTH CARE

Managing and coping with change

A study was undertaken, with colleagues at Imperial College between 1992 and 1994, to establish present practices and future requirements for senior management development and organisation development in District Health Authorities and NHS Trusts (Dawson *et al.*, 1995b). It is important to note that, since the study was completed, the structure and orientation of the NHS has once again changed to reflect the policies of the Labour Party, which was elected in 1997 (Sutherland and Dawson, 1998).

Field work was conducted in 21 sites and included interviews with 271 clinicians, general and functional managers who, at the time of the study, had substantial senior executive responsibilities as chief executives, executive directors, clinical directors and managerial heads of professional groups. The aim of the project was to establish what senior executives did, what problems they faced and their views on career development and learning needs. This section of the chapter deals only with the findings about approaches to managing change.

An organisational context was revealed in which senior executives felt themselves to be the subject (and, for some of them, the victim) of excessively high rates of organisational change, which in turn were associated with high rates of anxiety. Their accounts became familiar. Taking local as well as national initiatives into account, many respondents had lived through three or more major reorganisations. Every time there was reorganisation, they experienced massive dislocation. We were often told of circumstances in which managers, wishing to continue in what they saw as their job, found they were required formally to reapply for it when a new structure was established.

Many senior managers saw change as something with which they coped, rather than being something which they could create or steer or engineer. The language they used suggested they placed more emphasis on formal approaches to strategic planning and reacting to changes with which they felt bombarded, and less emphasis on developing their capacity for actively and pragmatically managing and creating sustained home-grown change. In terms of knowledge and learning, they 'knew' their 'old ways', they had 'learned' that change was something which they had to tolerate, that new initiatives were likely to be short lived and that their survival depended on their adjusting to whatever arrived on their desks 'for implementation'. Many of them had not found change to be a positive or cumulative experience which encouraged learning 'new ways' or building up knowledge which would underpin sustained improvements in performance.

Another aspect of management was that many respondents worked to limited managerial horizons of internal management and had not developed external aspects of their roles. This was particularly notable amongst the clinical professionals in senior management (Dawson et al., 1995). Within their complex and fast-changing environment, senior executives identified a variety of learning needs. They placed considerable emphasis on social skills and their need to develop capacity and competence in external managerial roles in marketing, communication and influence. Particular emphasis was placed on their role in managing relations between functions, professions and organisations.

There was widespread recognition that a major challenge facing senior managers in health was the need to learn how to improve their existing capacities to manage across boundaries. In NHS trusts there was much discussion about the need to manage more effectively across clinical disciplines to secure better patient-focused care. In district health authorities cross-boundary challenges were identified in terms of bringing public health, needs assessment, quality assurance and epidemiology together to inform purchasing and contracting decisions. In community care, boundaries between social services, housing, education and primary and community care groups were seen to inhibit more effective ways of working with the elderly and mentally ill.

In discussing the needs for management development, the importance of developing a complementary agenda for organisation development was stressed. For example, if cross-functional, cross-disciplinary and cross-organisational working is to be encouraged, managers need more support in addition to that which flows from their functional or departmental group. A further consideration is that, if managers are to become more outward looking, and seek to encourage the participation of diverse groups in advancing an agenda of performance improvement, their role in creating a common values framework will be vital in facilitating such participation and achieving improved performance.

In commenting on how they learned, and therefore on the sort of management development programmes that might be effective, senior managers stressed the personal importance they attached to 'learning by doing' though facilitated experience, mentoring and role models. Being apprenticed to someone who could demonstrate how to thrive on the complexities and uncertainties of health, and who could show how to manage outwards and across boundaries, was identified as invaluable in helping managers to perform well in senior positions in the NHS.

Conceptualising careers

In the same study, we examined the views of the 271 respondents on their career expectations. The findings reported above on the executives' feelings of being unable to seize the initiative and fashion change to suit their own circumstances, but rather being subject to changes made elsewhere, relate, at least in part, to their view of their own careers as fairly linear. One of two viewpoints was often cited. The majority of clinical specialists defined their career development in terms of an accumulation of professional qualifications and experience. In contrast, a minority of clinical experts and a majority of general and functional managers saw their career development as embedded in their experience of different organisational contexts. These findings revealed that few players in the health sector were thinking in terms of a third viewpoint, that of demonstrable managerial competence. Putting all three viewpoints together, we created a career climbing frame for the NHS. This is shown in Figure 2.3.

Using the climbing frame, individuals could be encouraged to think of their career development in three complementary ways, only one of which was tied to professional qualifications. With the help of mentors and advisors, senior executives could be encouraged to consider how to enhance their range and depth of competencies and gain experience of working in, and with, groups and organisations with different structures and cultures. This could be achieved through periods of secondment, participation in joint project teams, careful succession planning and programmed career advice.

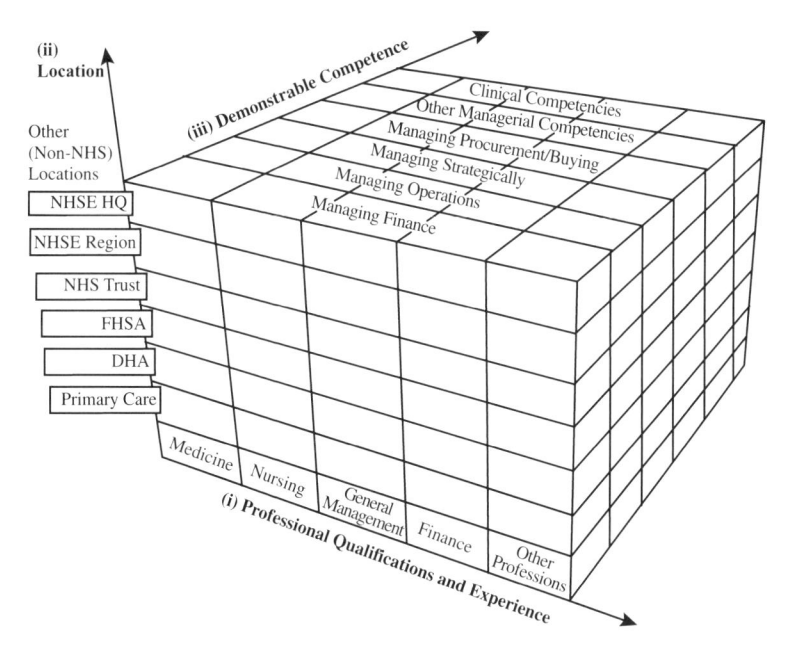

Figure 2.3 A career climbing frame for the NHS
Source: Dawson *et al.*, 1995a, fig. 3.1, p. 78

We concluded that such developments were very important and yet very difficult to achieve, for the following three reasons. First, they were important because strong strategic organisational needs for cross-boundary partnerships were identified as a prerequisite for improving performance in the determination of health priorities and the achievement of health objectives. Secondly, they were important because these aspects of management development are unlikely to be secured either by proclamation from on high or through reliance on organic evolutionary development. The emphasis placed on 'learning by doing' by senior executives means that direct experience is likely to be the most successful form of management development, albeit supported by more additional training and support. Lastly, they were unlikely to happen naturally because senior executives (whether initially managerially or medically trained) inhabited mind-sets which did not encourage them (because of concerns as various as competence, culture, status and pay) to think spontaneously and positively about crossing organisation, discipline or competence boundaries.

This discussion emphasises that management development must be considered in a context of organisation development. We recommended that NHS organisations develop an infrastructure which made career opportunities more visible to people from a variety of professional backgrounds and organisational experience, and, furthermore, that ways should be fostered to help people develop their careers and to provide bridges between individuals and organisations. Part of this would involve identifying the types of posts which are particularly suitable as 'sideways' moves between organisational locations, identifying opportunities for secondments and exchange between organisations and within organisations, between functions and service divisions, and supporting secondments, exchanges and cross-recruitment through human resource management practices, particularly appraisal, supervision and reward practices.

Managing and changing clinical practice

One of the aims of a more recent study of health management and organisation conducted with colleagues in the Universities of Oxford and Cambridge (Dawson *et al.*, 1998) was to understand the role of management and organisational interventions in securing change in clinical practice. The study involved field work in four hospitals and surrounding GP practices in 1995–7, including interviews with 203 doctors and 53 nurses involved in the diagnosis and treatment of two conditions, adult asthma and childhood glue ear.

In talking about their practice, doctors, just like managers, speak in terms of their own autobiography. They say they learn from their direct experiences: from patients, from colleagues, from 'seniors' to whom they consider themselves to be, or to have been, apprenticed. When they get to know of new developments, perhaps through being involved in the development or receipt of clinical guidelines or consensus statements, they are likely to consider how best these new ideas fit with what they do and know already. At the level of subjective understanding, doctors see their own clinical practice as beyond direct managerial intervention. They see the clinical domain as quite separate from the managerial domain.

However, when we looked beyond individual statements to patterns of responses we found that a significant part of individual biography is shaped by the organisational arrangements in which clinical practice occurs. Participation in multidisciplinary teams, the experience of leadership, and the development of communication systems with multiple cues to encourage behaviour change, are some of the things that emerged as having an effect on clinical practice. An interpretive understanding of what it means to be a clinician, of how clinical worlds of evidence and practice are created and sustained, enabled the research team to develop recommendations about further research into, and the practice of, changing clinical behaviour.

CONCLUDING COMMENTS: OBSERVATIONS ON THEORY AND PRACTICE

Three illustrations of research into aspects of work, organisation and management in the NHS have been given in order to demonstrated how field-work-based findings can be used to add to knowledge about the nature of management and organisation in health and to show that, within this knowledge, there are learning points for academics and practitioners in medicine and management. What then can we conclude for the theory and practice of health management? First, two sets of remarks on the practice of managing and organising.

Health systems are complex and contain enormous diversity. An action in one part of the system can have far-reaching consequences in others. Acknowledging and managing boundaries is a part of complexity. Professionally based training creates a number of tribes, each with its own assumptions and practices. Communication between tribes is often difficult. We see this in stark reality when looking at the relationship between the creators and disseminators of clinical research and clinical practitioners who have been described as inhabitants of different worlds (Dawson, 1997) or at the relationship between social workers, psychiatrists, psychotherapists, housing executives and community psychiatric nurses in the diagnosis, treatment and care of people who are mentally ill. It is a critical managerial task to find ways of enabling communication between the worlds, of creating synergy, but without losing the benefits of specialist knowledge.

A second practical conclusion is that, although health systems are vast and complex, individuals in roles as leaders, mentors, change agents and consolidators can make a real difference to the way in which local managerial and organisational systems are created and sustained.

The first step in being able to achieve change is for those involved to realise that change is possible. Managers and doctors are equally inclined to stress that they learn best by doing and through direct experience. Much of the work in health cannot be subject to mass standardisation or detailed hierarchical control. It needs to be customised to the context. Unless there is local ownership and commitment to solve the inevitable problems which arise, to train and motivate the people on the front line of action, to assemble appropriate resources and support, then, whatever a white paper or Act of Parliament decrees, health indices are unlikely to improve. Grand policies must be translated into meaningful action and meaningful lives at local level; and the role of leadership is critical in this translation.

Addressing now the social science research community, two conclusions can be drawn. The first is that, in order to achieve an accumulation of knowledge, research teams need to build on the hypothesis-generating capacity of case studies. We need carefully to consider how to conduct work at appropriate and different levels of analysis. Each of the studies which feature in this chapter has collected more data at the individual level of analysis than at the organisational. We have learned a great deal through the words and observed actions of individual clinicians or senior executives. Some data were collected at the organisational level and, indeed, in the first study we covered 21 locations. Nonetheless we have not yet developed sensitive and valid means to ensure strong and relevant organisational comparisons and we need to do this. Developing organisational metrics which relate to the structure, culture and systems of health organisations is a key priority.

A second challenge to the research community is to define and develop measures of a variety of content and process outcomes, so that one can undertake comparative work on the relationships between organisational and environmental contexts and aspects of performance in health care. Content indicators need to cover financial outcomes, clinical outcomes for patients following their interactions with the health service, and health status data for given populations. As we uncover more about the process of achieving effectiveness in organisations, we know we need also to consider process indicators. Three process indicators suggested in this chapter are capacity for inter-agency working, capacity for interprofessional working and capacity for generating and sustaining change, performance improvement and learning. Developing metrics for these is critically important.

A last conclusion for management researchers and management practitioners is that we should reflect on the way we can learn about management from the results of the second study concerning the relationship between clinical research and clinical practice. Parallels between the subject of this second study can be drawn to our endeavours to consider how organisational and managerial research relates to organisational and managerial practice. It shows that we need to understand the practitioners' world, to see it through their eyes. Only then can we hope to be able to develop a dialogue in which the findings of the research can enlighten practice.

Notes

1. I am indebted to Kim Sutherland for helpful comments on an earlier draft of this chapter.
2. This research project was funded by the Department of Health.
3. This research project was funded by NHS Executive North Thames, R&D Committee, Management and Organisation working group.
4. These estimates were given in an address in Cambridge in September 1997 on the subject of enterprise, by Dennis Gillings, the founder of Quintiles, a pharmaceuticals services corporation.

References

Burrell, G. and G. Morgan (1979) *Sociological Paradigms and Organisational Analysis* (London: Heinemann Educational Books Ltd).

Dawson, S. (1997) 'Inhabiting Different Worlds: How Can Research Relate to Practice?', *Quality in Health Care*, 6, 177–8.

Dawson, S., V. Mole, D. Winstanley and J. Sherval (1995a) 'Management, Competition and Professional Practice: Medicine and the Marketplace', *British Journal of Management*, 6, 169–81.

Dawson, S., D. Winstanley, V. Mole and J. Sherval (1995b) *Managing in the NHS: A Study of Senior Executives* (London: Publications Centre).

Dawson, S., K. Sutherland, S. Dopson, R. Miller (1998) *The Relationship between R and D and Clinical Practice in Primary and Secondary Care: Cases in Adult Asthma and Glue Ear in Children*, London, North Thames NHSE, R&D Committee.

Dionne, E.J. and J.S. Hacker (1997) 'Health Care Reform is Dead – Long Live Health Care Reform', *Annals of Emergency Medicine*, 30 December, 6, 742–5.

Entwistle, V.A., I.S. Watt, R. Bradbury and L.J. Pehl (1996) 'Media Coverage of the Child B Case', *British Medical Journal*, 22 June, 312(7046) 1587–91.

Lamm, R.D. (1994) 'Rationing and the Clinton Health Plan', *Journal of Medicine and Philosophy*, 19(5), 445–54.

Richmond, C., (1995) 'Is the Issue the Price of a Child's Life, or the Futility of Heroic Measures?', *Canadian Medical Association Journal*, 152(12) 2035–6.

Sellman, D. (1997) 'Child B: A Case of Just Care?', *European Journal of Cancer Care (English)*, 6 December, 4, 245–8.

Sutherland, K. and S. Dawson (1998) 'Power and quality improvement in the new NHS: the roles of doctors and managers', *Quality in Health Care*, 7, 516–23.

West, D.M., D. Heith and C. Goodwin (1996) 'Harry and Louise go to Washington: Political Advertising and Health Care Reform', *Journal of Health Politics, Policy and Law*, 21(1), 35–68.

3 Organisational Development: from a reactive to a proactive process

Peter Spurgeon

INTRODUCTION

This chapter will try to explore the concept of organisational development and, by implication, how the related concept of management development has operated in the NHS over the past decade. The basic proposition of the chapter is that organisational development has tended to become a largely reactive process attempting to implement, accommodate and at times ameliorate the impact of a range of externally driven policy initiatives. A consequence of this, it will be argued, is that organisational development informed by relevant theoretical and conceptual models is difficult to sustain. Instead initiatives aimed at shaping organisations can often be perceived as 'gimmicky' and transient, thus creating a negative and cynical audience.

The chapter will attempt to identify a set of forces that have created this situation and advocate a movement to a more proactive model of organisational development whereby organisations and staff working within them are properly developed and equipped to deal with reassessment and realignment of strategic direction and operational goals. Three main areas will be covered in preparing the argument: (a) the relationship being between policy, strategy and operational issues, (b) the growth of evidence-based medicine and its challenge to management, and (c) cultural aspects of organisations with respect to the issues such as leadership, team working and innovation.

POLICY, STRATEGY AND OPERATIONAL ISSUES

The process of health care system reform is widespread. Most countries have examined, or are examining, the functioning of their health care delivery system, initiating change with varying degrees of radicalism. Many authors have described and analysed this process (Tilley, 1993; Spurgeon, 1993) and it is not appropriate to cover this debate in depth here. Ham (1997) has summarised the main phases of reform, with the 1980s being described as concerned with macro-level cost containment, moving in the late 1980s to 1990s to efficiency and market-like mechanisms, while in the late 1990s more focused reform processes such as priority setting, managed care and evidence-based medicine are in operation.

In the UK in particular, the decade and a half of Conservative government saw reforms which, although inspired by the themes described by Ham, were also linked to a strongly dogmatic view of the inadequacy and inherent deficiencies of public sector provision. Policies were therefore characterised by an obsession with market-based mechanisms, with the inevitable competitive and divisive consequences as well as short-termism under the guise of destabilising large professional power bases and promoting greater efficiency by creating insecurity and fear of redundancy. It is perhaps not surprising, then, that many have observed the morale of NHS staff to be in crisis and that the new Labour government has focused on improved, ethically sound, management of staff as a key issue. Equally proponents of the private sector model of management might reasonably argue that the NHS has become much more responsive and customer-focused as a consequence of the reforms. However the merits or outcome of the reform are not our main concern here. It is rather to understand the way reform was attempted and the repercussions for organisational and management development.

The change process was for the most part centrally managed through a series of policies. As a consequence many senior individuals working in the NHS have become distracted by a need to analyse the latest policy statement and then try to estimate the next phase. There is here what could be described as 'distortion of policy' such that managers become ensnared in the glamour and intrigue of policy dissection and politicians misguidedly believe that the statement of policy is tantamount to the delivery of health care.

Endless policy analysis followed, much of which was commonsense interpretation and largely retrospective. Perhaps this was inevitable since there has been virtually no theoretical development in models of policy formulation since the 1970s. Thus the reform process proceeded surrounded by debate and argument, increasingly in the spotlight of the media and increasingly attracting chief executives of organisations into this arena. Almost as a natural compensatory consequence, chief executives became less concerned with day-to-day direct service delivery, preferring and being encouraged to focus on managing the political agenda. The dangers of this orientation are perhaps neatly summarised by Ham (1997), himself a policy analyst, who says: 'The evidence base of health policy is even weaker in many respects than the evidence base of medicine' and also that 'Real change in health care (from a patient perspective) derives as much from action by innovative managers and professionals as from initiatives taken by politicians'.

Therefore as the first point of my argument we see organisations and managers heavily engaged in an evidenceless, externally oriented debate when real change is more likely to be achieved by operational focus on direct service delivery. The most important danger, though, was of a gap developing between policy as espoused by politicians and managers and the sense of unreality amongst many NHS staff attempting to deliver this political agenda without resources or, in some cases, the necessary skills. One might wish to speculate whether some of the largest multinational private sector organisations would embark on fundamental change to the way they work without additional investment to support and sustain the process. Even with a concern for their profit margin, the answer is that they would not.

The market-based nature of the reform process compounded the tendency described above in reinforcing the need for managers to become more strategic, more externally focused (on the market) and to develop a vision for the future of the organisation. It is indisputable that strategic management has been a burgeoning area throughout the 1980s and 1990s. In terms of the instability in the external environment and globalisation of information, strategic management is clearly important, but it has acquired such an enhanced status that, almost by definition, operational aspects of management have been downgraded. Operational skills are frequently described as basic competencies, as if they are the practice of more junior staff who, if successful, will ultimately arrive at the strategic level.

Ironically perhaps operational tasks are where the majority of NHS staff are engaged. It is not the contention here that operational tasks are more important than strategic issues or indeed vice versa. It is that, through an over concern with policy (and politicians), a gap has developed between the two areas such that much of the workplace (operationally based) has become cynical and disillusioned about a policy agenda they feel powerless to influence while being vulnerable to its volatility.

Organisation and management development are themselves vulnerable processes in this environment. In attempting to support a turbulent external contact they become reactive and can become categorised as 'attempts to deal with the latest bright idea'. Perhaps no process better illustrates the disillusionment I have tried to describe than that of organisational downsizing. Embraced with gusto in the 1980s as producing 'lean and mean' organisations, it now transpires that this can lead to a stressed and overworked remaining workforce and a loss of important skills and experience for organisational regeneration. Observing this scenario is hardly likely to encourage staff to see processes of organisational change as helping them prepare to deal with the future. It also serves to weaken the notion of organisational development as a clear discipline with its own goals and objectives. In fact it appears as an attempt to cope with other forces, mainly financial, and therefore loses credibility in its own right. Essentially, then, our argument is that organisational development should move to establish principles with evidence of what an organisation ought to be like rather than what it is forced to be like.

EVIDENCE-BASED MEDICINE AND MANAGEMENT

The concept of evidence-based health care or evidence-based practice has taken hold in many countries. It has been grasped as a key concept of policy makers and is beginning to influence the organisation and delivery of services. The basic idea is clear and well known: health services should be based on the best evidence available and drawn from the findings of rigorously conducted research. As Walshe (1998) suggests, 'For too long patterns of clinical practice and the way in which we organise and deliver health care have been overly influenced by professional opinion, historical practice and precedent, clinical fashion and organisational and social culture.' However, as Walshe also notes, perhaps the greatest challenge in this whole area is not the accumulation of evidence and even its dissemination, but the changing of practice. Unfortunately it is in this area that the least evidence exists.

The weakness is highlighted as attempts are made to extend the notion of evidence-based practice to policy making and to management (Gray, 1997). The pressure to do this is clear and to some degree attractive. Many clinicians, their own practice under scrutiny, reasonably ask: what do managers do, how do we judge whether it is effective and how should the organisation be structured to be maximally effective? But the current evidence base for such judgements is very limited. Indeed it might be argued that the area of organisation and management development is not capable of being assessed in this way. If this is the case then organisational development as a discipline may be further undermined in that it appears to lack the conceptual or evidence base to advocate a position; that is, it is difficult to become proactive.

One of the repercussions of evidence-based medicine may be a challenge to management in terms both of the evidence available and of the nature and processes of research used to collect evidence. This may be an especially acute problem for organisational development given the complex and highly interactive environments in which our research is conducted. We may need to think carefully about the 'cult' of best practice studies which, although much vaunted by the current Labour government, do contain some significant conceptual limitations. In particular the following issues need to be addressed. Firstly the historical context of any initiative is largely ignored when best practice is described. The management of change literature, most notably Pettigrew, makes clear the role and impact of the historical background to a particular initiative. Many changes are in themselves a reaction or response to a previous situation. Even if researchers are able to properly document this previous context, it is virtually impossible for another organisation to replicate the sequence of events. This relates to the second point, which is the difficulty in establishing evidence of generalisability. From any case study of best practice it is typical practice to extract principles for dissemination and then implementation elsewhere. But these principles derive from a whole set of details and interactions which again are extremely difficult to describe and impossible to replicate. This is especially so with respect to individual personalities involved in organisational success. Champions of success is an oft quoted principle, but this is a description both of a role and of a specific individual. Moreover the personality dynamics may be crucial to the delivery of a particular outcome. For example, good practice in joint purchasing may be based on a positive dynamic between the chief executives of the health authority and the social services. How is a best practice process to be replicated if no such relationship exists?

The sustainability of effective processes is also largely ignored. Many successful models are a phenomenon of their time and appropriate behaviours in one time do not necessarily generalise to another. We are constantly told that health service managers must see change as the norm, that there is no stability – only continuous change. It is hard to see how in such an environment a static 'best practice' description can be appropriate for a new context that has already moved on. Fundamentally studies of best practice emphasise the best of local factors while failing to recognise that the next set of local factors may well be quite different.

The point here is that organisational development will remain reactive and vulnerable if it offers insight on the basis of transient, unstable evidence. If further evidence is unachievable then it is appropriate to fit the organisation around other forces as they may well have more justification for taking a particular direction.

CULTURAL ASPECTS OF ORGANISATIONAL DEVELOPMENT

The previous two propositions have focused on weaknesses or obstacles to organisational development emerging as a strong, proactive, shaping force. In terms of cultural aspects of organisations there is more room for optimism, but there are still issues to be clarified and oversimplification to be avoided. Cultural features are of course wide ranging and it is not possible to cover everything. By way of illustration concepts will be used relating to the contribution of leadership and to teamworking.

Definitions and constructions of culture abound. Bate (1984) has written widely on culture and argues that it is something that is implicit, that it is shared and may have an association with the way organisations attempt to solve problems. This positivistic notion was taken forward in the writings of Peters and Waterman (1984) who, in identifying successful organisations, seemed to endorse the somewhat simplistic notion that extension of the culture and practice of these firms elsewhere would also make other organisations successful. The seeds of 'best practice' studies may be observed here but, in the rather less successful future of some of the companies identified, so too can the defects of such studies.

There are some important lessons here for those working within the domain of organisational development. Identifying good supportive cultures and seeking to transfer such principles to other environments may well be overly simplistic. It also implicitly suggests that culture is largely defined by what is exposed by senior managers; that it is a 'top-down' driven process and that culture can be created and therefore manipulated. The cynicism that many staff feel towards mission statements would suggest that culture is something that needs to be developed co-operatively with all levels of the organisation. Indeed Alvesson and Berg (1992), in reviewing the evidence on the impact of culture upon performance, find it difficult to provide many examples of clear, unequivocal relationships. They point to the existence of many subcultures within organisations, local adaptation and the transitional nature of many of the effects as difficulties in seeing culture as a manipulable variable.

For example, there seems to be increasing emphasis in organisations on the need to encourage and foster innovation. If innovatory behaviour is truly distinct and different from normally adaptive problem solving then it is likely to be relatively rare and also likely to include a degree of risk of failure. There would seem to be some evidence (Anderson and West, 1992) that stability, continuity of membership and security engender more innovation through the enhanced ability to cope with the implications of risk. In the face of such evidence many organisations have spent a decade and more institutionalising insecurity by the relentless pursuit of short-term contracts and confirming the workforce as expendable in times of adversity. Much of the research literature on sources of stress in organisations would serve to support the growing perception of insecurity in the workforce (Spurgeon, 1998). The point here is that organisations need to behave in ways consistent with their stated goals and that those working in organisational development should be attempting to shape organisations to promote these desirable goals; that is, to be proactive, rather than to ameliorate the input of other external forces.

If culture and managerial behaviour are important forces in influencing organisational performance then we need to understand the directional and attributional aspects of these processes rather better. For example, in the context of leadership, Schroder (1984) has consistently argued that third world organisations with non-hierarchical, cross-functional structures will require a new set of competencies and skills to be utilised if they are to be managed effectively. This notion of new skills being required is explored by Spurgeon and Barwell (1990) in the context of organisational change. They argue, using the concept of cognitive maps or mental models, that the key to effective organisational management is the construction of mental representations of the way organisations function and interact, and how behaviours can support and facilitate the change objectives.

Two recent reports, Schroder and Spurgeon (in press) and Hackett and Spurgeon (1997), extend this theme of relevant behaviour for defined purpose. Much emphasis is placed upon teamworking, and the health service has been encouraged to develop a teamwork culture. This seems almost to be an indiscriminate advocacy irrespective of task or setting. Schroder and Spurgeon (in press) report a study based on a set of behavioural competencies in which utilisation of the competencies and hence associated performance are significantly better in the individual context rather than in a team situation. If teamworking is to be encouraged then a parallel process of learning how to work and be effective in teams seems to be necessary.

Hackett and Spurgeon (1997) point to the need for leadership behaviours to reflect the transformed nature of organisations such that leadership needs to demonstrate a recognition of the core values of the organisation and the consistent role of these values in determining behaviour, and to foster the participation and ownership of the workforce by the work contract.

In rehearsing these arguments in the three areas I have attempted (albeit selectively) to illustrate strands of ideas, concepts and evidence about how one might wish to see organisations develop in order to equip themselves and their staff to deal with future challenges. The challenge for organisational development is to have a clear commitment to creating such organisations in a proactive way.

What, then, is being advocated as proactive organisational development? The concept is similar to that involved in strategic management whereby proactive engagement with the issues is likely to be more effective (Asch and Bowman, 1989). Put into an organisational context, an organisational development initiative would be attempting to shape the organisation around the following principles:

- real and meaningful delegation (empowerment),
- a reduced distance between the hierarchy within any management structure,

- maximisation of shared meaning and expectations,
- an understanding of the implications of the process and focus of performance management,
- valuing continuity and security of providing supportive environments (as in the context of innovation and risk) and equipping existing staff with new skills for future requirements.

Obviously these are high-level principles and would need translation into organisational practice, but they are not 'soft' options. The effort and commitment to make organisations flexible and adaptive is more complex and more demanding than traditional models. For example, there would have to be considerable emphasis upon 'knowledge management'. Organisations cannot afford to lose the skills and experience of earlier situations. Our inability to learn for the future has tremendous social and individual cost.

Knowledge management would be a key to effective decision making, with evidence-based medicine seeing parallel evidence-based organisational development. The perhaps competing dimensions of rationality and creativity, and reflection versus action, would need to be brought together and made compatible. Some of this would need to be seen and symbolised by leaders capable of working in a variety of styles and settings. Organisations face a complex and demanding future. Organisational development has a role to join the pursuit of current operational efficiency with the desire for a smooth transition to the future. Unless this future shape is clear, and well defined, organisations will tend to lurch from one scenario to another, with painful repercussions for their own success and for the staff working in them.

References

Alvesson, M. and P.O. Berg (1992) *Corporate Culture and Organisational Symbols* (New York: De Guyters).

Anderson, N. and M.A. West (1992) 'Innovation, Cultural Values and the Management of Change in British Hospitals', *Work and Stress*, 6, pp. 293–310.

Asch, D. and C. Bowman (1989) *Readings in Strategic Management* (London: MacMillan).

Bate, P. (1984) 'The impact of organisational culture on approaches to organisational problem-solving', *Organisational Studies*, 5(1), 43–66.

Gray, J.A.M. (1997) *Evidence-based healthcare: how to make health policy and management decisions* (London: Churchill Livingstone).

Hackett, M. and P. Spurgeon (1997) 'Leadership and vision in the NHS: how do we create the "vision thing"?', *Health Manpower Management*, 22(1), 5–9.

Ham, C. (1997) 'Child's Play: Managing in today's National Health Service', Great Ormond Street Lecture, London.

Peters, T. and T. Waterman (1984) *In Search of Excellence* (New York: John Wiley).

Pettigrew, A. (1985) *The Awakening Giant: Continuity and Change in the Imperial Chemicals Industry* (Oxford: Blackwell).

Schroder, H. (1984) *Managerial Competencies* (Kendall and Dubuque, Iowa: Hunt Publishing).

Schroder, J. and Spurgeon, P. (in press) *Differential Performance between Male and Female Managers on High Level Management Competencies.*

Spurgeon, P. (1993) 'Regulation or free market for the NHS: A case for existence', in I. Tilley, (ed.), *Managing the Internal Market* (London: Paul Chapman Publishing).

Spurgeon, P. (1998) 'Managing stress in healthcare environments', in P. Spurgeon (ed.), *The New Face of the NHS*, 2nd edn (London: Royal Society of Medicine).

Spurgeon, P. and F. Barwell (1990) *Managing Change in the NHS* (London: Chapman & Hall).

Tilley, I. (ed.) (1993) *Managing the Internal Market* (London: Paul Chapman Publishing).

Walshe, K. (1998) 'Evidence-based practice: a new era in healthcare?', in P. Spurgeon (ed.), *The New Face of the NHS*, 2nd edn (London: Royal Society of Medicine).

4 Methodology and Marketing: making organisational behaviour research irresistible in health care

Edward Peck and Jenny Secker

INTRODUCTION

Although the organisational behaviour issues which need to be addressed within health care are well understood and the theoretical frameworks within which to analyse them are well established, the volume of organisational behaviour research undertaken in UK-based health care is small and the impact on local and national decision making appears to be negligible.

This is a disappointing state of affairs for any organisation, but especially so for the UK health service, which is the host for such a large volume of clinical research and which increasingly stresses the importance both of other research and evaluation activities and of evidence-based medicine. We do not intend to imply, of course, that managers are not interested in techniques that might improve organisational effectiveness. The recent trend in UK health care, however, is to bring in consultants as agents of organisational analysis and development, and those consultants often base diagnosis and intervention on approaches which are typically fashionable but atheoretical (benchmarking, for instance). Nonetheless most managers in UK health care are aware of the benefits of research and have an interest in organisational development. There are a number of obstacles, however, which appear to prevent managers from commissioning or facilitating organisational behaviour research, obstacles which the academic community may wish to address.

In the first section of this chapter we highlight three such obstacles: the demands imposed on health care managers' time by current techniques for data collection; a lack of 'fit' between research time scales and the fast-moving world of health care organisations; and the difficulties involved in disseminating research in ways which have relevance for managers themselves. We also provide illustrations of some possible ways around these obstacles drawn from recent research at the Centre for Mental Health Services Development (CMHSD). However the study from which we draw our illustrations throws into relief the potential for tensions between the pragmatic world of health care management and the quality and integrity of research. The second section of our chapter therefore examines these tensions from the perspective of recent work aimed at developing quality criteria for qualitative research. In conclusion, we outline the parameters of the debates in which we need to engage in order to negotiate them.

OBSTACLES AND WAYS AROUND THEM

In 1996, CMHSD was commissioned to undertake a research project on behalf of the King's Fund London Commission (Peck *et al.*, 1997). The study aimed to obtain descriptive data about the management of London's mental health service alongside an account of the concepts, behaviours and perceptions of the people responsible for managing the service system. The intention was to broaden the analysis of problems within mental health services in London beyond the collection of statistical and documentary data concerning demography, activity, service models and so on (see Johnson *et al.*, 1997). Funding for the research was secured by arguing that the report to the London Commission could not recommend particular interventions, such as increased funding or new policy requirements, unless the impact of these interventions was understood in terms of management capacity and capability to act on them. Examining the ways in which managers would be likely to interpret and respond to changes was therefore a central aim of the research. This study is used here to illustrate the ways in which we attempted to overcome the three obstacles identified above.

Demands on time

Typically, current techniques of data collection involve either postal questionnaires, used with large samples, or face-to-face interviews, with smaller samples. While the first approach attempts to minimise disruption to managers, it tends to produce poor response rates and little depth. On the other hand, although face-to-face interviews can overcome these problems, they generally make far greater demands on managers' time.

For our study, we attempted to find a way around this dilemma by combining a relatively brief face-to-face interview with an innovative approach to the collection of more detailed qualitative data. Before beginning the interview proper, managers were asked to use a Dictaphone to dictate an uninterrupted narrative of recent personal experience in developing a local mental health project. In order to stimulate participants' thoughts, an introductory letter had suggested some broad categories within which they might wish to construct their narrative. These included the nature/origins/sources of the change/innovation; the drivers for the change; obstacles; and lessons learnt for the future. We assumed that, as senior managers, participants would be familiar with using a Dictaphone, and the majority of participants were indeed comfortable with this approach, dictating narratives which lasted on average around 20 minutes. Once they had completed their narrative, participants were interviewed using a structured questionnaire which also included a small number of open questions.

All 57 managers who were asked to take part in the research agreed to do so. Arguably the relative brevity of the combined dictation and interview was a key factor in enabling us to achieve this response rate. Beyond a commitment not to make unrealistic demands on managers' time, however, we also thought it important to deploy four of our most senior staff to undertake the field work. In addition to the fact that the researchers were therefore known to most participants, thus facilitating access to their diaries, this meant that they shared the social and intellectual background of the managers they were interviewing. The researchers themselves also had considerable knowledge of mental health service policy, practice and management. This approach is uncommon in organisational research, where the field work is usually conducted by research assistants unfamiliar with the participants and/or with the topic area. Despite the cost implications for the funders, we believe the approach had advantages, to which we will return later.

Differences in time scale

It is widely recognised that qualitative research in particular is a time-consuming exercise, not only as a result of the length of interview usually required, but also because the process of analysis involves continual movement between the data and emerging themes in order to adapt and verify the analytical framework being developed. This lengthy process contrasts sharply with the time scales within which health care managers must often operate and may mean research findings are no longer relevant by the time they become available.

For our study of London managers, we needed to find ways around this obstacle, not least because the research received funding in June 1996, with a deadline for the final report of September 1996. In more typical research studies, the design phase alone might take longer than the three months this allowed.

As an organisation undertaking both consultancy and research, CMHSD is familiar with consultancy projects with tight deadlines dictated by the fast-moving policy and practice environment within which health care managers operate. However we were concerned that to attempt to conduct the study within this time scale would be to sacrifice quality to haste. In order to deliver on time, we prioritised the project, switched resources from consultancy work and bought in additional staff to collaborate in the field research, develop a database and input data. Nonetheless the analysis of participants' narratives still posed a challenge in meeting the time scale and we therefore needed to find ways of short-circuiting this process. The method we adopted involved using a 'shorthand' version of the 'frameworks' approach to qualitative data analysis described by Ritchie and Spenser (1994). Rather than developing the main themes of the analysis from the transcripts in the kind of iterative process outlined above, the four interviewers agreed a coding framework on the basis of notes they had made during the dictation of the narratives. Each researcher then searched their quota of the transcripts for data reflecting these themes, recorded their presence and selected quotations which illustrated the point being made. To ensure that we did not omit important data, we also included any topics lying outside the main themes we had identified which were of particular significance to the participants.

Having identified key themes in this way, we held a discussion session in order to establish points of similarity and dissimilarity between participants from different agencies. We also used this session to identify themes that had emerged from the analysis of both the questionnaires and the transcripts. Arguably the method produced a rich source of data from which to identify common underlying phenomena influencing the management of mental health care, explore (un)common decision-making processes and map the range of contributory elements that influenced service management and, in turn, service delivery. Certainly, of all the material collected for the King's Fund London Commission (Johnson et al., 1997), this study appears to have had a significant influence on the chapter on mental health in the final report of the Commission (King's Fund London Commission, 1997), putting an organisational behaviour perspective at the heart of the discussion.

Routes of dissemination

Typically OB research is disseminated through academic publications and chapters which people not familiar with the academic repertoire can find difficult to follow. Where health service managers are concerned, normal dissemination routes may therefore be inaccessible, particularly since research currently being undertaken at CMHSD confirms anecdotal evidence that managers have little time for reading. This has two major consequences. Firstly, managers are much less likely to seek out research-based evidence if their experience suggests that this will not be presented in a manner which is accessible. Secondly, managers will not be weaned off their present attachment to the kind of atheoretical fashions discussed earlier unless research findings are presented in an equally accessible way.

In order to overcome these obstacles, the study described above was brought to the attention of health care managers through a variety of routes. In addition to the press launch of *London's Mental Health* by the King's Fund, we ensured that the *Health Service Journal* was aware of the interest that the chapter reporting our study would hold for managers. This led to a specific news feature on that chapter (Crail, 1997). Together with the King's Fund, CMHSD also hosted two one-day feedback sessions for mental health service managers in London to brief them on the contents of *London's Mental Health* and seek their responses. These included enthusiasm for a mental health development network which would adopt an overtly theoretical approach to the local exploration and resolution of the system pathologies identified by our study. The network was established across three localities in September 1997. Finally, the findings were presented at a number of conferences and seminars throughout 1997.

This is not to suggest that we do not value academic routes for dissemination. On the contrary, like other university-based researchers, we have an obligation to contribute to the UK government's Research Assessment Exercise, and in any case we are aware of the mutual benefit of presenting our work for peer review and scrutiny. Our original research, now augmented by a follow-up study undertaken in the summer of 1997, has recently been submitted for publication. However we would argue that the primary audience for organisational behaviour research should be managers, not academic colleagues, and for this reason we concentrated our first efforts on that audience.

PRAGMATISM OR QUALITY?

While we would argue that the study described above yielded findings of interest to both the academic and practice communities, we are aware that our approach was not always consistent with current thinking about the quality of qualitative research. In recent years, health care researchers have expressed concern both about the difficulties of publishing qualitative work in a field which is still dominated by the positivist tradition and about the standard of some of the work which is published. In an attempt to address these issues, criteria have been put forward to assist researchers and reviewers in ensuring and assessing the quality of qualitative studies (Popay *et al.*, 1998; Secker *et al.*, 1995). Here we draw on this work to highlight the potential for tensions with the study used to illustrate the previous section.

A starting point for the establishment of the quality criteria which have been put forward is the location of qualitative research in the epistemological tradition of interpretivism, associated primarily with anthropology and sociology. As its name implies, the central tenet of this tradition is that we inevitably interpret everything we observe and experience through a mesh of understanding, or 'sieve', woven from our previous experiences and interpretations of information from other people and sources. From a research perspective, this has three important implications. Firstly, it follows that the purpose of research is not to establish objective 'facts' about the social world, because objective knowledge is impossible. Rather the aim is to explore how research participants understand, or make sense of, the topics in which we are interested. Secondly, the theories we arrive at as researchers are also inevitably our own interpretations of research participants' understandings, and not simply a reflection of them. Finally, in order for readers and reviewers to assess qualitative research, it is necessary to present a 'thick description' of the data (Geertz, 1973), to illustrate this with substantial extracts, and to make the processes of the analysis transparent (Secker, *et al.*, 1995).

Taking the first implication, the exploration of research participants' understandings is not a straightforward matter of asking questions and recording answers. Rather the researcher has to minimise the intrusion of his or her own preconceptions by setting aside prior assumptions and exploring in depth how participants' responses make sense in the context of their own biography, social circumstances and culture (Garfinkel, 1967). This is regarded as particularly important in 'insider' research, where researchers and participants share a similar practice background, because the risk of unwarranted assumptions of shared meaning is greater. One solution put forward by Garfinkel (1967) is for 'insider' researchers to treat accounts of situations and issues which appear familiar as 'anthropologically strange'. By this Garfinkel means that researchers should not assume that their usual understanding of familiar situations is adequate, but should treat the situations as though they were unfamiliar in order to reveal the usually unremarked ways in which they are constructed.

In practice, however, this would mean lengthy research interviews in which health care managers may be unable or unwilling to participate. As has been seen, we were able to engage managers in qualitative research by using narratives recorded by Dictaphone without interruption on the part of the researcher, and hence without any further exploration. Clearly this raises the question of whether we were really able to interpret these recordings accurately, or whether we have simply jumped to conclusions on the basis of our own assumptions about mental health services management.

Turning to the second implication of interpretivism, researchers who work within this tradition acknowledge that the theories they generate through qualitative research are themselves interpretations of research participants' understandings. For this reason emphasis is placed on ways of ensuring as good a 'fit' between the two as possible. These include the need for the researcher to become 'immersed' in data in order to generate analytical categories from the data themselves (Glaser and Strauss, 1971). As categories and theories linking them emerge, these must be continually checked and verified against the data in an iterative process of constant comparison analysis (Ritchie and Spenser, 1994). Again this raises the question of whether we may have drawn conclusions based on unwarranted assumptions, or on selective and therefore potentially misleading data, by short-circuiting the process of analysis in order to meet the required deadline.

Finally the need to provide 'thick descriptions' of data means research reports must go beyond merely stating 'facts' independent of intentions and circumstances. Rather the context of participants' perceptions and experiences must be described, together with their intentions and meanings (Popay *et al.*, 1998). Equally making the analytical process transparent involves providing evidence of the steps taken to ensure that all the available data have been accounted for. As a minimum, Rogers and her colleagues propose that it should be possible to decipher the analysis from descriptions of the research setting and interactions, and from participants' accounts. Given the length of extracts often necessary to enable readers to make sense of qualitative data, this makes for substantial reports which can be difficult to accommodate even within the word limits required by academic journals.

As has been seen, in disseminating our study we used more accessible routes, such as the *Health Service Journal*, feedback sessions and conferences. However length restrictions meant that we were unable to illustrate either the process of analysis or the data themselves in any detail. By implication, then, we expected health service managers to accept and act on our findings in the absence of the level of evidence normally expected of research reports.

DISCUSSION

If the tensions we have outlined above are to be successfully negotiated, we believe it is necessary to engage in debate about a number of key issues. Central to these is the question raised earlier about 'insider' research and about whether, in some cases, the 'fit' between researchers' and participants' backgrounds may be so close that less in-depth probing is necessary. As has been seen, the research team who undertook our study of London managers themselves had extensive experience of managing mental health services. In these circumstances, in-depth probing might have placed unreasonable demands on managers' time while yielding little further useful information. It is also likely that we would not have achieved such extensive coverage of the sample, since managers might have been reluctant to agree to a longer interview.

In this context, our use of dictated narratives appears to be a new technique. It may be in the development of such techniques, which enable managers to participate in research without distracting them from their management task, that further work is needed. As a further example, we are currently developing a structured participant diary to assist in the evaluation of the impact of consultancy interventions concerning mental health and primary care.

Equally, where the process of analysis is concerned, it may not always be necessary to generate every analytical category from the data themselves when the researchers and participants share a similar background. As a further illustration, many of the analytical categories we are currently using in a study of managers' professional development needs are derived directly from the research questions, since these address issues such as supervision and personal development planning which are as familiar to the researcher, herself a service manager, as to participants. This is not to imply that participants' experiences within each category are identical. Clearly these experiences require more detailed analysis using the constant comparison method outlined earlier, as do particular sections of the interviews where a critical incident technique was used to explore participants' experiences of successes and failures. In addition we have found it necessary initially to keep an open category for coding data which do not readily fit any preconceived category. Nevertheless we have been able to short-circuit aspects of the painstaking process of analysis by using categories derived from the interview schedule.

Finally, where dissemination is concerned, it is arguable that a wide range of routes and repertoires is required if research is to have the intended impact on service development. We would therefore propose that speedy dissemination in the more accessible journals and through conference presentations is acceptable, providing that a full report on the basis of which the quality of the research can be judged is made available as soon as possible. We are also currently assessing the value of the Internet as a means, not only of disseminating research findings, but also of inviting responses so that we can reflect on our conclusions in the light of alternative interpretations.

CONCLUSION: IMPLICATIONS FOR THE FUTURE OF ORGANISATIONAL BEHAVIOUR RESEARCH IN HEALTH CARE

Our experience of research and consultancy in the field of mental health services management has convinced us that there is an increasing need for qualitative research undertaken within time scales which fit more closely those within which managers themselves operate. In developing the techniques required to achieve this, we would argue that greater attention could be paid to technologies, such as the use of Dictaphones and the Internet, on which managers themselves rely to meet deadlines in the fast-moving world in which they work. A first implication, then, is that researchers will need both to acquaint themselves with these technologies and to keep up to date with new developments which could be adapted for the purposes of research.

However we are aware that managers are not the only consumers of organisational behaviour research and we would not suggest that the kind of solutions we have proposed obviate the need for longer-term qualitative research, particularly in less well conceptualised areas. For example, we are currently developing work aimed at exploring the relationship between aspects of health care organisation and outcomes for patients, and we would not propose that this could be achieved within the limited time scale of our London managers' study. A further implication, then, is the need to develop clear criteria, driven not simply by funders' time scales but also by our own research agenda, as to when it is appropriate to short-circuit the processes of qualitative research and when it is not.

In turn it will also be necessary to develop quality criteria for qualitative research which reflect the varied contexts within which different studies are undertaken. One way forward here would be to test empirically the question we have raised about whether 'insider' researchers can short-circuit the processes of exploration and analysis, or whether this results in jumping to conclusions on the basis of false assumptions about shared meanings. For example, where results are required within a short time scale, as with our work for the King's Fund London Commission, it might be possible to meet the required deadline but to set aside time for more detailed enquiry and analysis at a later date, thus allowing for comparison of and reflection on the conclusions reached.

For the present, however, we believe the study we have described deepened managers' own understanding of the service system in which they were working, strengthened their interest in adopting more theoretical frameworks for organisational development and, albeit more tentatively, influenced the national policy agenda. These would appear to be appropriate aims for organisational behaviour research in health care. If quality criteria for qualitative research could be developed to reflect such aims without compromising integrity, organisational behaviour research might become irresistible to its principal target audience.

References

Crail, M. (1997) 'Urban blight', *Health Service Journal*, 23 January, 12–13.
Garfinkel, H. (1967) *Studies in Ethnomethodology* (Englewood Cliffs, NJ: Prentice-Hall).
Geertz, C. (1973) *The Interpretation of Cultures: Selected Essays* (New York: Basic Books).
Glaser, B. and A. Strauss (1971) *The Discovery of Grounded Theory: Strategies for Qualitative Research* (Chicago: Aldine/Atherton).
Johnson, S., R. Ramsay, G. Thornicroft, L. Brooks, P. Lelliot, E. Peck, H. Smith, D. Chisholm, B. Audini, M. Knapp and D. Goldberg (eds) (1997) *London's Mental Health: A Report to the King's Fund London Commission* (London: King's Fund).
King's Fund London Commission (1997) *Transforming Health in London* (London: King's Fund).
Peck, E., H. Smith, I. Barker and G. Henderson (1997) 'The obstacles to and opportunities for the development of mental health services in London: the perceptions of managers', in S. Johnson *et al.* (eds), *London's Mental Health* (London: King's Fund).
Popay, J., A. Rogers and G. Williams (1998) 'Rationale and standards for the systematic review of qualitative literature in health services research', *Qualitative Health Research*, 8(3), 341–51.
Ritchie, J. and E. Spenser (1994) 'Qualitative data analysis for applied policy research' in A. Bryman and R. Burgess (eds), *Analysing Qualitative Data* (London: Routledge).
Secker, J., E. Wimbush., J. Watson and K. Milburn (1995) 'Qualitative methods in health promotion research: some criteria for quality', *Health Education Journal*, 54(1), 74–87.

5 The Consumer's Role in Co-ordination: making sense of transitions in health care

Alexandra Harrison, Amy Pablo and Marja Verhoef

INTRODUCTION

There is an international trend towards integrated health care delivery systems (Lomas, 1996; WHO, 1996). An integrated delivery system is a network of organisations that provides or arranges to provide a co-ordinated continuum of services to a defined population and is willing to be held clinically and fiscally responsible for the outcomes and the health status of the population served (Shortell *et al.*, 1996). In Canada health regions were created in Alberta, in June 1994 by merging previously independent organisations responsible for acute care, continuing care, home care and public health under a single board with a single administrative structure. Health regions in Alberta exhibit many of the features of integrated delivery systems. There has been a consolidation of hospital services, which is an example of horizontal integration – the co-ordination of functions, activities or operating units that are at the same stage in the process of delivering services (Gilles *et al.*, 1993). More significantly regions also include vertical integration – co-ordinating, linking or incorporating activities or entities at different stages of the process of producing and delivering care (Conrad, 1993) which is evident in the organisational linkages between acute care and community-based services.

This organisational restructuring has significant implications for organisational behaviour. Kaluzny and Shortell (1994) predict a major change from managing an organisation to managing a network of services or a continuum of care with an increased emphasis on managing across boundaries. Although a health region is a single organisation, the historical interactions between previously independent organisations, particularly those that provide different phases of care, mean the transitions between these sectors of a single organisation need to be explicitly managed in order to achieve real integration.

Co-ordination is one means of linking different organisations or various parts of a single organisation (Longest and Klingensmith, 1994) and is an intended and anticipated outcome of the new regional health care system in Alberta: 'The Regional Health Authorities Act will promote co-ordination and integration of health services' (Alberta Health, 1994). Co-ordination has traditionally been viewed as a critical organisational element, both in the organisational literature (Galbraith, 1973; Ring and Van de Ven, 1992) and in the health services literature (Alter and Hage, 1993). There are, however, very few empirical studies about co-ordination (Alter and Hage, 1993; Bolland and Wilson, 1994; Devers *et al.*, 1994) and these studies do not examine the consumers' view. When consumers have been consulted about co-ordination (Grusky and Tierney, 1989), the focus has remained on administrative procedures, with no indication of whether consumers considered this to be important.

In addition to providing new challenges for health care managers, navigating the transitions in restructured organisations has implications for those who are cared for, as well as those providing the care. This chapter reports important aspects of the findings from a study that examined consumers' views about co-ordination and their experience of transition from one sector of the health care system to another. The findings emerged as part of a qualitative study conducted in the Calgary Health Region in Alberta, Canada. The main purpose of the study was to develop a substantive grounded theory (Glaser and Strauss, 1967) about the meaning of co-ordination to consumers who had experienced the transition from an acute care hospital into the Home Care Program in the Calgary Regional Health Authority. This chapter explores a key aspect of the findings: the consumers' role in co-ordinating their own care. The results are analysed from the perspective of sense making, using the properties described by Weick (1995). The intent is to enhance our understanding of consumers' experiences as they make sense of their transition between sectors in the health care system.

METHOD

Setting

Alberta is one of ten Canadian provinces, which have constitutional jurisdiction over the delivery of health services. Alberta covers an area of 255 000 square miles, with a population of over two and a half million people. In June 1994, the Regional Health Authorities Act was proclaimed which established provincial regions for the delivery of health services.

The Calgary Regional Health Authority (CRHA) is one of 17 regional health authorities created. The region provides primary and secondary care for citizens in the Calgary area (about 800 000 people), tertiary care for surrounding regions and certain quaternary services for the entire province. The region has a single regional board (replacing seven previous boards), one regional medical staff and one regional senior management structure with a single chief executive officer.

Participants

The informants for this study were consumers who had experienced the transition from an acute care hospital in Calgary into the Short Term Home Care Program. Short-term clients are an important group of home care clients. They make up about 20 per cent of the total number of clients but account for more than 50 per cent of professional home care services. One of the rationales for regionalisation is to shift the site of care from hospital into the community and short term home care clients are the group that are most likely to reflect this change. In Alberta, as in the rest of Canada, the length of hospital stays has been decreasing. The Short Term Home Care population is the group that is most likely to include the patients that have experienced early or timely discharge from hospital.

Design

The study used a qualitative, grounded theory approach (Guba and Lincoln, 1994). Qualitative research involves the collection, integration and synthesis of non-numerical narrative data (Leasure and Allen, 1995). The appropriateness of the qualitative approach is derived from the nature of the social phenomenona to be explored (Morgan and Smircich, 1980). Qualitative methods are characterised by the search for meaning in particular settings in which human beings are considered active forces (Jones, 1988). This was an exploratory qualitative study, since the variables are unknown, the context is important and there is no established theory base for generating hypotheses that could be tested quantitatively (Creswell, 1994).

Data collection and analysis

An interview guide was used with open, unstructured questions asking about (a) the person's experience leaving hospital and (b) the meaning of co-ordination to the consumer. More structured prompts were introduced as the interviews progressed to elicit information about categories that began to emerge as important in the transition experience. With the permission of the interviewee, all personal interviews were recorded and later transcribed. An iterative, constant comparison process was used to analyse the data (Strauss and Corbin, 1990). The transcripts of the interviews were reviewed and portions of the text were identified with a particular word or phrase (a code) that captured the meaning of the passage. With further analysis, various codes were related to each other, and this linking process developed into an index system. Nud*ist software was used to record codes and categories in the index system, as well as the relevant sections of the interviews. Richards and Richards (1994) describe this as the code and retrieve method and observe that the generation of categories through analysing the data is an important contribution to developing theory. In writing up the findings there is a continual cycle of reviewing emerging categories as new sections of text and memos from the researcher are added, summarising the text in each category and revising the index system so that there is congruence between the categories, the sub-categories and the individual codes. As the theory develops, portions of interviews are used to document why particular codes were chosen, as well as to support the emerging theory.

FINDINGS

A total of 33 consumers who were recently discharged from an acute care hospital and then received short-term home care services were interviewed, including 26 consumers who participated in an in-depth personal interview and seven who were interviewed by telephone. The participants were similar in age and sex to the entire group of consumers who were eligible for the study. However some of the eligible consumers could not be interviewed because of limited English-speaking ability or cognitive impairment.

A striking aspect of the study that emerged from the interviews is the importance of consumers in co-ordinating their own care. This came up in more than half of the interviews. It was a factor for men and women, and for all age groups. Consumer involvement meant that the consumer took particular actions in a variety of domains. These action roles – communicate, monitor, manage and direct – are described below and illustrated with quotations from consumers.

Communicate

There were two aspects to the communicate role: the first involved the consumer interacting with various care providers, the second was an active consumer role in seeking information, often from sources other than health care providers.

In the following example, as well as communicating with each care provider, the consumer is a pivotal communication link between the care providers. The Home Care nurses would call the patient to ask how often her doctor wanted her dressings changed. The nurses would then alter the schedule of home visits accordingly.

Consumer 14 (female, age 50)

INTERVIEWER   *Do the* [Home Care] *nurses call the doctor's office or does the doctor call the nurses?*
CONSUMER   *I think he gets hold of them but it's usually done through me first.*
INTERVIEWER   *So he tells you about the changes?*
CONSUMER   *Yes. When he changed it* [the number of times the dressing should be changed] *from twice a day to once a day, the nurse that initially came, the one that usually comes in, phoned me that afternoon after I'd been to the doctor and asked me what he said. Then she changed it* [her home visits] *down to once a day.*
INTERVIEWER   *Okay, so the nurse basically did it through you?*
CONSUMER   *Yes. Through me.*

The other aspect of the 'communicate' role was consumers gathering information to inform themselves from whatever sources they had available. In the following example, the consumer sought information from her friends regarding what to expect before going into hospital or coming home.

Consumer 9 (female, age 79)

I did quite a bit myself before I went into hospital. I phoned people that had been in the———Hospital or had been in surgery and I asked them what I was supposed to take and my clothes. And another thing is that I asked them everything that I could about what, like after care or things like that. And they told me.

Monitor

In the 'monitor' role, consumers take on more responsibility related to their care. In the first example below the consumer indicated it was up to her to carry on with her therapy and keep track of how she was doing.

Consumer 15 (female, age 82)

She checked my exercises. She hasn't had a chance really to do much. I'm just carrying on with the exercises the nurse in the hospital gave me. And I keep a chart for myself to see if I'm going up or going down or what I'm doing.

A more active monitoring role was evident with another consumer who wanted to avoid problems. She had a midline intravenous line inserted while in hospital and it was very clear to her that the hospital staff had no experience with this.

Consumer 4 (female, age 35)

There were no nurses on my floor that knew anything about a midline at all. In fact when they put it in, I had several nurses come in to watch them put it in because they had never seen it and were interested in seeing how it was done. It was like the talk of the floor because nobody really had seen it before.

When the hospital nurses wanted to start using the midline intravenous line the consumer became involved in this decision.

Well, actually there was a point where the regular intravenous line had come out and the nurses [in the hospital] *wanted to start using the midline – because I hadn't started using it, they* [Home Care] *had just inserted it – and I didn't want them to touch it because I was afraid they would do something. So I had them phone the Home Care nurse and see how to do it first. A couple of the nurses weren't very happy with me but at that point I thought – it took them forever to put the midline in, leave me alone. That was a little tense.*

Manage

There is a further evolution of consumer involvement with the manage role. In the next example the consumer was engaged in adapting his environment both in the hospital and at home to allow him to manage with severe injuries to one side of his body.

Consumer 5 (male, age 39)

CONSUMER   *I had to call my brother to bring in proper duct tape and make one* [modifications to his walker] *in the hospital. That same walker is still in the hospital on reserve for me whenever I go in there for physio.... The occupational therapist introduced me to various things that should help around the house. None of them worked. For example, she was trying to get me to use a tub seat and a bar bolted to your tub to allow you to get into the tub. I told her that it just wouldn't work but she tried to get me using it. What would be involved was I would have to use my left arm to support myself to do the mechanics in the bathroom. My left arm is the one that is damaged* [shattered in a car accident] *that I can't give support to.*
INTERVIEWER   *Have they been able to offer any suggestions or equipment that you might use, or ideas about how you can basically fix it up for yourself?*
CONSUMER   *I've basically kind of fixed up most of it myself.... I think what it comes down to is who's the co-ordinator? The co-ordinator seems to be me, the customer.*

Direct

The final step in the evolution of the consumer role is 'direct'. In this example, an ex-navy man was very definite that, to make things happen in the health care system, either in hospital, or at home, or in the transition between the two, a person had to take charge themselves.

Consumer 8 (male, age 76)

INTERVIEWER *What would you tell your buddy about the experience of being in hospital and coming home?*
CONSUMER *Don't leave it to them. Take your situation in your own hands. You have to take your situation in your own hands.*

DISCUSSION

Transitions involve change and uncertainty. In hospital, the environment is unfamiliar for the consumer but the health care is provided in a relatively structured and controlled manner, with the opportunity for interaction and validation with a variety of care providers. The role of the consumer is often seen as relatively passive, possibly owing to the unfamiliarity of the environment and the dominance of the care providers. When the consumer goes home, however, the home environment is more familiar, the health care providers are not continuously present and the care provided is less predictable.

In this study, many consumers indicated that they played an active role in the co-ordination of care, in assisting the transition between hospital and home, and in participating in the care they received at home. Consumers undertook four actions as part of this role, with increasing levels of consumer control with the various roles identified. There is a progression from the passive aspects of the 'communicate' role along a continuum to the very dominant role the consumer played in the 'direct' role. These consumer actions are examined using the concept of sense making as a framework for developing a greater understanding of the consumers' views of their transition experiences.

Sense making

Weick (1979) suggests that 'believing is seeing' since people make sense of things by seeing the world in a manner that is consistent with their beliefs about that world. Sense making involves putting stimuli into some kind of framework that helps people to understand and to predict. To make sense of things, people 'read into things the meanings they wish to see; they vest objects, utterances, actions and so forth with subjective meanings, which helps to make their world intelligible to 'themselves' (Frost and Morgan, 1983, p.207). Weick (1995) stresses that sense making should be distinguished from interpretive activities, such as interpretation and metaphor (Morgan, 1986), since it is a literal, not an interpretive activity: it is what it says – making sense. Weick (1995) identifies various properties of the sense-making process which are applied to analysing the findings from this study.

The 'communicate' role: a social activity involving cues

It is enlightening to consider the 'communicate' role in the context of two sense-making properties described by Weick (1995): 'social' and 'cues'. Conduct is contingent on the conduct of others, whether those others are imagined or physically present. Sense making is never solitary because what a person does internally is contingent on others. This social property of sense making is evident in the first dialogue that was used to illustrate the 'communicate' role. In that example, the consumer plays the central role in a series of conversations involving the consumer, the nurses and the doctor.

Communication by and with the consumer, as well as involving a social interaction also influences cues. Extracted cues are based on the familiar aspects of a situation and serve as seeds that allow people to develop a larger sense of what may be occurring. The context is important for determining what the extracted cue becomes, first because of the frame or structure of the context, and secondly by affecting the interpretation of the cue. Cues are an important reference point for sense making since they tie elements together cognitively (Weick, 1995).

An aspect that emerged from the interviews as important to consumers in helping them prepare for the transition home was knowing what to expect. Cues may be an important element of building expectations because cues direct what a person focuses on. The importance of social interaction and cueing was evident in the second case used to illustrate the 'communicate' role. This woman called her friends before she went to hospital to ask them what to expect, to help her prepare for her experience.

The 'monitor' role: retrospective, ongoing and driven by plausibility

Understanding the 'monitor' role is enhanced by considering the three characteristics listed above which are central to the sense-making process. Weick (1995) argues that plausibility is more important to sense making than accuracy. This is because people need to be selective in what they attend to because there are so many data available that it is easy to become overwhelmed. Stimuli that detract from a response are often filtered out. People see and make sense of things they can do something about. The capacity for action, rather than accuracy, affects what is believed or rejected. Sense making involves plausibility, coherence and reasonableness. In the findings reflecting the 'monitor' role, the consumer identified a plausible explanation for the fact that the home therapist did not seem to be helping much with the consumer's home exercise programme. Her explanation of what was occurring involved actions that she could take herself; that is monitoring her own progress.

It has been stressed that the most distinguishing characteristic of sense making is the retrospective focus (Weick, 1995). People can only know what they are doing after they have done it. The creation of meaning requires directing attention to events that have already occurred, backward from a specific point in time (now). Thus whatever is occurring in the present will influence what people discover when they look back. The other example used to illustrate the 'monitor' role was the woman who ended up having difficulty with a special type of intravenous line that had been inserted so that she could receive treatment at home. Her experience exemplifies the importance of the retrospective focus. She was told to go to the emergency room at her community hospital to deal with the problem, but no one there knew what to do. For this woman, her assessment of what happened was determined in retrospect, since at the time it was 'confusing'. As she looked back, she indicated that she 'spent hours in Emergency and came out of there thinking I could have done this myself'.

A final sense-making category that is related to the concept of 'retrospective' is the property of 'ongoing', which means sense making does not suddenly start or stop. When people are involved in an organised sequence of actions and are interrupted, they try to make sense of it. If the interruption slows the accomplishment of the existing sequence of events, people are likely to experience anger. In the example above, the consumer expected the problem to be dealt with promptly at the Emergency Department. When this did not occur she was very irritated about it.

If, on the other hand, the interruption accelerates accomplishment, then the person is likely to experience pleasure. Another consumer in the home IV programme was able to get out of hospital much sooner and was, therefore, very pleased with his experience. The emotions generated by interruptions in this 'ongoing' property affect sense making because past events are reconstructed in the same emotional feel as the present (Weick, 1995).

The 'manage' role: identity construction

A consumer who had been severely injured in a car accident assumed the 'manage' role to tailor his environment. This was a way to re-establish a positive self-image and make sense of both his care experiences and, more importantly, his new life experience. Part of identity construction is projecting the new identity onto the environment and observing the consequences. The consequences for this man's new identity were positive. He was able to do something better than the professionals who were supposed to be knowledgeable and his expertise was acknowledged.

An important aspect of identity construction as part of sense making is that it is self-referential. It may be the self, rather than the environment, that needs interpretation. What the situation means is affected by the identity adopted in dealing with it, which in turn is related to what one thinks is occurring (Weick, 1995).

The 'direct' role: enacting the environment

Weick (1995) uses the term enactment to highlight the point that people are part of their environments and often actually produce aspects of the environment they deal with. As people act, they create constraints and opportunities in the environment. This leads to a self-fulfilling prophecy: people create and find what they expect to find since they act in such a way that their assumptions about reality are in fact fulfilled.

One of the roles that consumers assumed in this study was to 'direct', which involved enacting their environment. For example, the ex-navy man quoted in the findings was very definite that, to make things happen, a person had to take charge. This consumer had been hospitalised on many previous occasions for a variety of conditions. His success in enacting his environment may have been related to his past experience with the health care system. His other life experiences in the navy and as a salesman may also have helped him believe that he could influence his environment, which is how he behaved, which, in turn, produced the results he expected.

Summary

The analytical framework provided by the sense-making perspective enhances our understanding of the way consumers make sense of a transition experience between two sectors of health care. Once people begin to act (enactment) they generate tangible outcomes (cues) in some context (social) and this helps them discover (retrospect) what is occurring (ongoing), what needs to be explained (plausibility) and what should be done next (identity enhancement) (Weick, 1995, p.55). Transitions require a 'communicate' role which involves social activities and cues. The 'monitor' role reflects the retrospective, ongoing nature of sense making which is driven by plausibility rather than accuracy. The 'manage' role allows identity construction and the direct role is a way to enact the environment.

Changing roles

The findings identified a number of actions that consumers undertook in co-ordinating their care. A review of Weick's seven properties of sense making adds insight about these actions from the consumers' perspective. This active role that consumers play in the co-ordination of their care has implications for the organisational role of consumers, the roles of professionals and the culture of health service organisations.

Mills and Moberg (1982) identify a number of differences in processes and outcomes between organisations which manufacture goods and those which produce services. A major difference in service organisations, such as those that provide health services, is that the customer and the service provider must interact. An implication of this interaction is that, as well as providing the information that constitutes the 'raw material' for the interaction, service organisations also often make use of clients' efforts in actually producing the service. For example, the customer fills in a deposit slip in the bank, or patients carry their x-rays from one department to another in a hospital. However consumers may have less knowledge than they require to carry out the tasks they need to do. Thus they may need to be co-opted into organisational membership roles so they can acquire the necessary information and use it responsibly.

Patient satisfaction with clinical care is increasingly considered as outcome in its own right (Harrigan, 1992) but consumers of services also have important role activities to perform in the production of the services they use. To become active participants, consumers have to acquire the knowledge, skills and interest that will enable them to perform as effective, although temporary, members of service organisations (Mills, 1986). This could evolve into an implied contract between the consumer and the organisation. However, before an implied contract can develop, there needs to be a more explicit recognition of the consumer's organisational role.

This study suggests that, in addition to assessing the outcomes of care, consumers can and should be involved in the processes of delivering that care, particularly in a vertically integrated system where co-ordination between sectors is a goal of the system. Patients and their families are eyewitnesses to the process of health care delivery (Gerteis *et al.*, 1993). Since consumers are the stakeholders who experience the entire episode of an illness, they are ideally positioned to participate in, or to evaluate, the continuum of care. Active consumer involvement, as participants who are involved in co-ordinating their own care, is one mechanism for accomplishing the intra-organisational, inter-sectoral boundary spanning that is required to achieve an integrated system. Thus the consumer becomes an integral part of the organisational system, central to the integration, enabling the production of service and quality desired by the consumer (Mills and Moberg, 1982).

For consumers to participate as partners in their co-ordination role they need assistance and support from health care providers, with professionals serving as teachers or coaches as well as providing health care. Supporting consumers to allow them to fulfil the roles identified means providing information (often in a variety of formats), answering questions, giving advice about how to manage potential problems and, most importantly, providing an atmosphere in which consumers feel their contributions will be valued. The role of the consumer in coordinating their care needs to be explicitly recognised by health care professionals and health service organisations. Co-optation of consumers into organisational membership roles (Mills and Moberg, 1982) may be necessary to ensure that important information is shared and may be the mechanism through which their cultural influence is legitimised and institutionalised.

CONCLUSION

The findings from this study have implications for the future research agenda of organisational behaviour in health care including the organisational roles of consumers, the roles of health care managers and providers and the culture of health service organisations.

Mills and Moberg (1982) suggest that the consumer, as a partial organisational member, deserves more organisational research. Specifically the design and socialisation of the consumer role has received little attention. Although this study provides some information about the role of 'consumer' in the co-ordination of services, there are still many questions about how to formalise these roles and socialise consumers, providers and managers to play and support these consumer roles.

The important consumer role in an integrated delivery system has implications for health care providers since they need to act as teachers and guides as well as clinicians. A further challenge for organisational behaviour research is to identify and evaluate what other competencies will be required for health care providers and managers as organisational forms, such as integrated delivery systems, emerge in response to environmental or legislative changes.

There was considerable internal validity demonstrated for the findings with the group of consumers who were interviewed. However the findings need to be validated with other groups of consumers, such as those who could not be interviewed because of limited English-speaking ability or cognitive impairment. In this study consumers were leaving acute care, an environment of consumer dependence with high levels of professional care provided, to go home – an environment with more consumer independence and more limited amounts of professional care. It would be useful to examine whether the role of the consumer is different for other transitions, such as moving from acute care to continuing long-term care. This transition would involve institution-based care for people on both sides of the transition which might affect the consumer role.

A consistent theme in the properties Weick identifies for sense making is that of a self-fulfilling prophecy as individuals interact with their environment. This is most obvious for 'enactment' but is also evident in 'identity construction', 'social', 'cues' and 'plausibility'. During the present study, the environment was very turbulent. The newly formed health regions were dealing with significant reductions in funding (15–20 per cent), concurrent with massive restructuring of health service organisations. It would be interesting to investigate whether the strong consumer role found in this study is evident in a more stable environment, considering the strong environmental influence in Weick's view of sense making.

With the current trend in health care to vertical integration of services, dealing with boundaries and transition processes across both internal and external organisational boundaries is an important issue for health care organisations. Consumer involvement may be a key, not only for consumers making sense of their experience, but as an organisational strategy to help manage transition processes.

References

Alberta Health (1994) *Regional Health Authorities User's Guide* (Edmonton, Alta: Alberta Health).
Alter, C. and J. Hage (1993) *Organizations Working Together* (Newbury Park, Calif.: Sage).
Bolland, J. and J. Wilson (1994) 'Three faces of coordination: A model of interorganizational relations in community-based health and human services', *Health Services Research*, 29(3), August, 341–66.
Conrad, D.A. (1993) 'Coordinating patient care services in regional health systems: The challenge of clinical integration', *Hospital and Health Services Administration*, 38(4), winter, 491–508.
Creswell, J. (1994) *Research Design, Quantitative and Qualitative Approaches* (Thousand Oaks, Calif.: Sage).
Devers, K., S. Shortell, R. Gilles, D. Anderson, J. Mitchell and K. Erickson (1994) 'Implementing organized delivery systems: An integration scorecard', *Health Care Management Review*, 19(3), 7–20.
Frost, P.J. and G. Morgan (1983) 'Symbols and sensemaking: The realization of a framework', in L.R. Pondy, P.J. Frost and T.C. Dandridge (eds), *Organizational Symbolism* (Greenwich, Cohr.: JAI).
Galbraith, J. (1973) *Designing Complex Organizations* (Reading, Mass.: Addison-Wesley).
Gerteis, M., S. Edgman-Levitan, J. Daley and T. Delbanco (1995) *Through the Patient's Eyes: understanding and promoting patient-centered care* (San Francisco: Jossey-Bass).
Gilles, R., S. Shortell, D. Anderson, J. Mitchell and K. Morgan (1993) 'Conceptualizing and measuring integration: findings from the health systems integration study', *Hospital and Health Services Administration*, 38(4), winter, 467–89.
Glaser, B.G. and A.L. Strauss (1967) *The Discovery of Grounded Theory: Strategies for Qualitative Research* (Hawthorne, NY: Aldine de Gruyter).
Grusky, O. and K. Tierney (1989) 'Evaluating the effectiveness of countywide mental health systems', *Community Mental Health Journal*, 25(1) spring, 3–20.
Guba, E.G. and Y.S. Lincoln (1994) 'Competing paradigms in qualitative research', in N.K. Denzin and Y.S. Lincoln (eds), *Handbook of Qualitative Research* (Thousand Oaks, Calif.: Sage).
Harrigan, M.L. (1992) 'Outcome evaluation', *Quality of Care: Issues and Challenges* (Ottawa: Canadian Medical Association).

Jones, M.O. (1988) 'In search of meaning: using qualitative methods in research and application', in M.O. Jones, Z. Mocre and R.C. Snyder (eds), *Inside Organizations: understanding the human dimension* (Beverly Hills, Calif.: Sage).

Kaluzny, A. and S. Shortell (1994) 'Creating and Managing the Future', in S.M. Shortell and A.D. Kaluzny (eds), *Health Care Management, Organization Design and Behaviour* (Albany, NY: Delmar Publishers).

Leasure, R. and P. Allen (1995) 'Introduction to the research process', in L.A. Talbot (ed.), *Principles and Practice of Nursing Research* (Boston: Mosby).

Lomas, J. (1996) 'Devolved Authorities in Canada: The new site of health-care system conflict?', in J.L. Dorland and S.M. Davis (eds), *How Many Roads: Regionalisation and Decentralization in Health Care* (Kingston, Ontario: Queen's University Press).

Longest, B. and J. Klingensmith (1994) 'Coordination and communication', in S.M. Shortell and A.D. Kaluzny (eds), *Health Care Management, Organization Design and Behaviour* (Albany, NY: Delmar Publishers).

Mills, P. (1986) *Managing Service Industries* (Cambridge, Mass.: Ballinger).

Mills, P. and D. Moberg (1982) 'Perspectives on the technology of service operations', *Academy of Management Review*, 7, 467–78.

Morgan, G. (1986) *Images of Organization* (Newbury Park, Calif.: Sage).

Morgan, G. and L. Smircich (1980) 'The Case for Qualitative Research', *Academy of Management Review*, 5(4), 491–500.

Richards, T. and L. Richards (1994) 'Using Computers in Qualitative Research', in N.K. Denzin and Y.S. Lincoln (eds), *Handbook of Qualitative Research* (Thousand Oaks, Calif.: Sage).

Ring, P.S. and A.H. Van de Ven (1992) 'Structuring cooperative relationships between organizations', *Strategic Management Journal*, 13, 483–98.

Shortell, S., R. Gilles, D. Anderson, K. Erickson and J. Mitchel (1996) *Remaking Health Care in America: building organized delivery systems* (San Francisco: Jossey-Bass).

Strauss, A. and J. Corbin (1990) *Basics of Qualitative Research: Grounded Theory Procedures and Techniques* (Newbury Park, Calif.: Sage).

Weick, K.E. (1979) *The Social Psychology of Organizing*, 2nd edn (Reading, Mass.: Addison-Wesley).

Weick, K.E. (1995) *Sensemaking in Organizations* (Thousand Oaks, Calif.: Sage).

World Health Organization (1996) *European Health Care Reforms* (New York: World Health Organization).

6 Professional Control Issues between Medicine and Nursing in Primary Care

Lynn Ashburner and Katherine Birch

INTRODUCTION

There are issues of professional control implicit in the policy moves towards a 'primary care-led NHS'. As the primary care sector becomes the focus for NHS service developments, policy initiatives continue to be guided by this concept (NHS (Primary Care) Act 1997 and the more recent white paper, *The New NHS – Modern and Dependable*, 1997). Recent policies such as the introduction of GP commissioning groups, have put the GP at the centre of all primary care services and not just those traditionally attached to general practice. However the role of nursing, having been ignored in the past, is acknowledged in the white paper. For each new primary care group, there is to be a nurse representative on each board, although the actual significance of this needs to be carefully assessed. Such recent trends have increased the power and control doctors have over other health professionals, but, as independent practitioners, GPs may have little experience of working in a corporate way organisationally, and equally may have little experience of management and co-ordination. These additional roles will add to the already growing number of new responsibilities. A key issue is the relationship between doctors and other health professionals over 'contested boundaries'. These can be discussed in terms both of actual tasks performed and of who has control over the tasks.

A focus for current attention is the development of advanced nursing practice in primary care and the issue of their substitution for GPs. This chapter will present findings from a number of studies on the role of GPs and nurse practitioners, with a focus on a scoping exercise carried out by the authors on the incidence and role of nurse practitioners in primary care (Ashburner *et al.*, 1997).

In this chapter the term 'primary care' is used to refer to all non-acute care, whether provided within a hospital setting, by a GP or in the community. This comprises the great majority of all NHS care. The NHS, particularly primary care, is currently facing pressure from a number of sources. The public demand for health care continually increases, and developments in needs assessment have identified populations whose health needs are unmet. Furthermore, as some services are transferred from acute settings into the community (such as care of the mentally ill, care of the elderly and minor surgery) and as hospital stays become shorter and the number of beds decrease, the role of primary care services is both crucial and central. The nature of much of this work falls within the existing responsibility of nursing groups, but 'boundary' issues will still occur as and when duties previously carried out by GPs, *whether or not they actually require a medical practitioner*, are transferred to nurses. In this context, direct 'substitution' of nurses for GPs should be seen as a possible long-term objective rather than the 'answer' to current service problems.

In the long history of interprofessional power struggles over boundaries between the medical profession and nursing, nurses have experienced both gains and losses (Davies, 1980). At a time when there is this a great increase in service need and the potential for growth in nursing's contribution to primary care service delivery, it is important that issues of control are addressed and also that they do not impede innovation and development. This is important, not just from an autonomy or professional perspective, but also to ensure that the most appropriate modes of health care delivery are developed in response to health needs.

ISSUES OF PROFESSIONAL CONTROL

The health sector presents a valuable focus for studies of professional behaviour in organisations given the relative importance of professional power when compared with non-public sector organisations. The professions and boundary issues in health care can be conceptualised as follows:

- the development of individual professions and their struggles to define and control their boundaries;

- the strategies and struggles over boundaries between professions for contested terrain;
- the boundaries between professional and outside groups where power is contested – this would include management;
- the definition of boundaries in relation to a changing environment – policy changes, social expectations, technological development, changes in the mode of delivery, and so on.

The history of the NHS can be seen in the above terms, with government playing a significant role in its backing of certain groups, most notably the medical profession.

The key concepts are 'professional dominance', the collective dominance of one professional group over other groups, including the patient, and 'professional autonomy', which refers to the different aspects of control the group has, in terms of political autonomy, economic autonomy and technical or clinical autonomy. Professional autonomy issues are prevalent in the debate between managerialism and professionalism (Ashburner, 1996), but the concept of professional dominance and theories of occupational closure are those which are most appropriate in examining the relationship between medicine and nursing. The functionalist perspective (Parsons, 1951) emphasises the differences and separateness between professional and non-professional occupational groups, whilst that of the interactionists (Abbott, 1988) offers greater scope for understanding changing boundaries between groups by defining the professions as just one group within an occupational category.

The concept of professional dominance refers to various forms of dominance not only over other health occupations but over the patient and social definitions of ill-health and so on. Freidson's critique of medical power (1970) questioned the basis for this dominance, noting that it should not be seen as 'a benign form of social control' but as power without accountability, and as such should be subject to regulation by the state. Medicine's gains have been matched by nursing's losses. Salvage (1988) refers to nursing as the 'failed profession' (p.517), and even the inroads into the autonomy of different groups of health professionals, made by both Griffiths (1983) and the 1990 Health and Community Care Act, have had a greater impact on nurses than on doctors (Elston, 1991).

Both professional groups will develop their own strategies for defending or expanding boundary issues. Historically the medical profession's dominance has seldom been severely challenged by nursing, and the relevance here is that it is necessary to understand whether this is still going to be the case given the pressures on GPs and the relevance of the nursing role for primary care. For nurses, the attempt at reducing the level of professional dominance of the medical profession could lead to their greater autonomy.

This raises several significant questions regarding independent practice and autonomy. Given that in the short term what is being addressed is the potential transfer of non-medical tasks to nurses, without an increase in autonomy, one would imagine that this type of change would be unchallenged, from a political, professional and economic perspective. So why are so many doctors resistant to it? Do they see it as the thin edge of the wedge? Do they see it as making nursing autonomy more likely? Why do some resist allowing nurses greater scope or autonomy, even when it is one solution to their problems of overwork and recruitment? The issue of who might be the most appropriate person to deliver care in terms of patient need or health need does not seem to be important in the factors taken into consideration by doctors. Collectively issues of professional medical control appear to be of key importance in the maintenance of professional dominance.

NURSING AND MEDICAL ROLES IN PRIMARY CARE

Over the last decade there has been a remarkable growth in the numbers, scope and role of practice nurses and the emergence of the 'new' category of the 'nurse practitioner'. This term was developed in the USA to refer to independent nursing practice, but this is not the case in the UK, where there is no agreed definition of the level or the scope of the work that nurse practitioners can or should do. Thus the development of the role has been ad hoc and is largely the outcome of an increased GP workload following the new contract (Department of Health, 1990). In consequence, the way that practice nurses are developing their role relates predominantly to the particular needs and views of individual GPs rather than to patient or health needs, or within a framework for professional development. Nurses in general practice are employed directly by GPs, so it is they who control training and the definition of nursing duties. As Harrison and Pollitt note (1994, p.4):

> The medical profession buttresses its own position through widespread involvement in the training and state registration of other

health professionals, based upon a widely accepted claim that medical knowledge is all-encompassing of health services, other professions therefore being logically subordinate.

There are fundamental differences between the medical model and the nursing model of care as far as understanding of illness is concerned. The medical model sees illness as an individual pathology, biologically grounded, rather than one which is socially, economically or politically created. If doctors dominate emerging definitions of nursing work, this may serve to limit the potential for the holistic nature of nursing developments. Beyond individual practices, there are other groups of nurses with a greater level of autonomy, such as district nurses and health visitors. If the GP is to lead and manage primary care services, their autonomy is at risk, as such roles may in the future cease to exist or be subsumed into new structures.

Within the UK, historical definitions of GP and nursing roles have produced an imbalance between the roles, and research has shown that GPs can spend a substantial amount of their time on duties that do not require a medically qualified clinician (Honigsbaum, 1985). Given the increase in the non-medical workload resulting from the new GP contract (in 1990) and its requirements for increased screening, for example their involvement in purchasing, and changes in the configuration of services, GPs are facing increasing pressures. The number of medical students choosing a career in general practice is declining and it is estimated that this will be exacerbated by the recent changes in undergraduate medical education (Calman, 1993).

The growing pressure on primary care provision is a phenomenon shared by many developed and developing countries and the extension of the nursing role is one possible response. It has been argued that nurses who combine traditional medical skills such as diagnosis, with nursing skills, such as nursing assessment, are able to provide a different and more comprehensive service to patients than that which is provided by medical practitioners (Fawcett-Hennessey, 1991; Jackson, 1995). The former service is typically regarded as more flexible, combining both 'caring' and 'curing' approaches (Salvage, 1985) and may be more suited to the needs of patients who receive care within the primary care sector and nurses may be the most appropriate providers for a large proportion of this care. Since health care in the community has traditionally been associated with long-term care, supporting the disabled and chronically ill and assessment of patients' long-term health and social needs, there could be opportunities for nurses to play a leading role in these areas.

In this context the differences between the 'medical' and 'nursing' models of care take on greater significance, with the holistic nature of nursing care appearing to be the more appropriate, despite the dominance of the medical model throughout the NHS. As Harrison and Pollitt (1994) say, 'the very clear, individually-oriented model of ill-health, sometimes referred to as the "medical model", [is the one] upon which the NHS is constructed' (p.5).

The development of nursing roles within primary care may be regarded as an efficient and effective means to meet the demands currently facing primary care services. Looking to the experience of nurses in America, the role of the nurse practitioner (an advanced clinical nurse specialist, with specific qualifications) has come to be regarded as central to the provision of health services across a broad range of patient groups and in a variety of settings, but being predominant in primary care (Safreit, 1992; Mundinger, 1994; Jordan, 1994). Some nurse practitioners in the USA have independent practices with complete autonomy, whilst others work in GP clinics or within the acute sector. Statutory prescribing powers have also been given to such practitioners, on a state-by-state basis. Evaluations of nurse practitioners in the USA indicate that the services that they provide are clinically equivalent to those provided by general practitioners, and patients typically respond positively to such services (Brown and Grimes, 1995). Indeed, the Department of Health itself acknowledges this:

> Research has shown that nurses can be as effective as doctors – and as acceptable to patients – in securing compliance with therapy for chronic disease, making initial assessments of patients, diagnosing and treating certain minor acute illness and behavioural disorders and rehabilitating elderly patients after surgery. (Department of Health, 1996, p.31)

While the US system of health care provision differs from that in the UK, it is apparent that the situation in the USA is not unique, particularly when we consider the nature of professional autonomy, boundaries and clinical freedom. Doctors in the USA felt 'increasingly threatened by the enhanced role and correspondingly improved status of nurses' (Iglehart, 1987) and such conflict over interprofessional roles manifested itself in a number of ways. The American Medical Association, for example, withdrew from a scheme to improve the educational and political status of nurses on the grounds that it was against the best interests of medicine. It is not clear, however, 'whether the argument was based on the effectiveness of patient care or the relative status of physicians and nurses' (McKee and Lessof, 1992).

Given the pressures which primary care services in the UK are currently facing, the roles of doctors and nurses are undergoing considerable change and opportunities exist for the respective roles of medicine and nursing to be reviewed and for nurses to achieve a greater degree of autonomy, whether under the title of 'nurse practitioner' or not. The white paper (Department of Health, 1997) also provides for the inclusion of nursing at a senior level in commissioning. The challenge for both professional groups is to avoid the conflicts that have occurred in the USA. The relationship of nurses to other health care professionals also needs to be considered, together with associated issues such as prescribing behaviour, which itself requires legislative change. Health services as they exist today mainly reflect the professional interests and concerns of the medical profession. The new arrangements envisage a greater influence for the professional interests and concerns of the nurse, although whether such change can be realised in practice depends on a number of factors.

ISSUES AND EVIDENCE

The emphasis on primary care, the GP role and the changes to service delivery raise a number of important issues relating to workforce planning, professional development and service provision, which go beyond those covered in this chapter. For example, if greater control over primary care commissioning is to be given to GPs, what happens to, amongst other things, the public health and needs assessment roles, currently the responsibility of health authorities? To what extent will other key health professionals either retain control over the scope of their work or be in a position to influence their own professional development?

The past decade has witnessed several attempts by the nursing profession to 'shake off the shackles of medical dominance by getting government support for measures designed to enhance its professional independence' (Salter and Snee, 1997, p.30). Within nursing, the development of a pay review body similar to that in medicine and clinical nurse grading, reflect the demands of nurses to be recognised as a profession. In 1992, the UKCC (United Kingdom Central Council) introduced the Scope of Professional Practice and Code of Professional Conduct which, 'in conjunction with the new system of post-registration education and practice, jettisoned the doctor-dependent, extended role procedure and proclaimed that each nurse was responsible for maintaining and developing their own competence and practice' (Salter and Snee, 1997, p.30). While such moves reflect the desire amongst the nursing profession for greater autonomy, it is apparent that the scope for nursing to achieve greater control depends upon more than professional guidelines. Recent research suggests that advances in the scope and practice of nursing remain closely linked with medical opinion as to the need for and desirability of such roles and the issue of where legal liability lies (Ashburner *et al.*, 1997).

In 1996, the NHS Executive (R&D Directorate, West Midlands Region) commissioned a research project designed to gather information on the nurse practitioner role. In addition to an extensive literature review in the USA and the UK, all health authorities in England were asked to provide data relating to three broad themes: the prevalence of local nursing innovations in primary care (specifically the development of nurse practitioners), the training that such nurses had received and the demand amongst GPs for increasing the role of nursing within primary care. A 70 per cent response rate was achieved for this project, the results of which are incorporated into the discussion below.

At a policy level, the issue of workforce planning has concentrated on the possibility of 'substituting' nurses for doctors. The concept of 'nurse substitution' is relevant for longer-term health care planning, but problems connected with what has been perceived in some quarters as a direct challenge to medical power, concerned with issues of realignment and boundaries, have arisen. This should not deter the move to increase nursing roles within existing parameters.

Within the changing nature of primary care provision, the relationship between a changing health care policy agenda and nursing practice is both complex and diverse. Despite primary care practitioners being described as the 'crest of health services', it is argued that the potential of primary health care services is not being realised owing to outmoded methods of working and specialisation and structures which fragment the primary care service (Quinney and Pearson, 1996). Such criticisms include the traditional dichotomy between medical and nursing practices, whereby nursing has been perceived of, and has developed, as the 'handmaiden' of the medical profession (Salvage, 1990). The factors implicated in shaping new roles and in changes to existing roles in primary care nursing are therefore extremely complex, incorporating national initiatives, local pressures and overcoming traditional professional boundaries and control. Despite developments occurring within primary care nursing, particularly at the level of the practice nurse, it is apparent that the extension of the practice nurse role was predominantly related to the changing workload and responsibilities of GPs. Over the last ten years there has been an extraordinary growth in the number of practice nurses (Rowley, 1994) and in the range of services they provide (Atkin and Lunt, 1995; Hunter and MacAuley, 1996). The necessity for this proliferation has been attributed to the growing demands placed upon general practitioners to provide health promotion and maintenance services through the new GP contract and the 1990 Health and Community Care Act (Rowley, 1994) and a growing recognition that a large proportion of the work that GPs undertake does not make full or appropriate use of their skills and knowledge. Since GPs are reimbursed up to 70 per cent where they can demonstrate that practice nurses are fulfilling GP's obligations for health promotional and screening activity, much of practice nurses' activity is concerned with this area. Within this context, the cost agenda of the GP or practice, as opposed to quality or health need issues, may determine the boundaries of nursing activity.

If nursing is to respond to such diverse pressures it needs to speak in a unified voice as a collective, as does the medical profession. There are several reasons why this has proved a problem in the past, and one key consequence has been the lack of a framework for or any coherence in the professional development of nursing. A significant finding from the national study of nurse practitioners and innovations in primary care nursing was the wide range of titles currently in use to describe nurses working within primary care settings, none of which defined clearly what role might be expected of the individual nurse. There was a general lack of a clear statement as to how such roles had been defined and developed, in terms of either the qualifications or the training of the individual involved. In describing the tasks that nurses within primary care were undertaking, it was apparent that there was the potential for a high degree of overlap to occur between nurses with differing titles. Conversely nurses with the same title may be involved in contrasting areas of service provision, depending upon the GP practice. This diversity of title and role reflects the fractured way in which primary care services have developed:

> It is likely that nurses, doctors and government are interested in the concept of the nurse practitioner for different reasons. Nurses wish to extend their role (Greenfield *et al.*, 1987), doctors want to share their workload and enhance the practice team and the government is probably looking for a cheaper way of meeting health service needs. (Sailsbury and Tettersell, 1988, p.316)

Another influence on such developments, at a practice level, came from health authorities where the predominant concern was to purchase the most appropriate and cost-effective forms of health care delivery. This context of looking to the longer term and to wider issues might result in their objectives being aligned with those of nursing.

Although only three GPs were interviewed for the project, they were selected on the basis of their being at the 'leading edge' of promoting the use of nurse practitioners in primary care. While all stated that this would give nurses greater autonomy, in fact they were talking about nurses carrying out tasks without direct supervision but still very much under the control of the GP. Hence the GP's use of the concept is not commensurate with common usage. There was no view expressed, for example, that nurses should become involved in diagnosis. The notion of autonomy as used by the GPs related to nurses' ability to carry out GP-designated tasks unaided, and the ability to 'fit in' was judged to be *far* more important than any specific clinical skill. This does not suggest a style of working whereby nurses could gain any degree of independence or autonomy. Similarly Atkin and Lunt (1995, p.5) describe GPs as being 'particularly attracted to the idea of having direct access to, and control over, a nurse situated in the practice'.

The expanding role of the GP does not just involve continuing control within the practice but also threatens the existing high levels of autonomy of nurses who work in the community, such as health visitors and district nurses, as GPs become the organising centre of primary care. Traditionally health visitors and district nurses have practised independently, within their own nursing hierarchy. Current policy initiatives which describe the GP as being the 'hub' of all primary care services implicitly question the viability of these two types of community nurse.

DISCUSSION

Within primary care, a high degree of power has been handed to GPs. This may have serious implications for the role of nursing within primary care. It is apparent that, while the nursing profession itself is seeking to develop roles and further nursing autonomy, particularly within primary care, GPs still retain a high degree of control over the tasks that nurses undertake, and are unwilling to promote autonomous nursing practice. Doctors' continuing concerns over issues relating to dominance, even if on the face of it they have little to gain, may relate to their need for a strong base for their power, from which they continue to fight issues of autonomy.

The many strands of managerialism as they have been introduced into the NHS have challenged medical dominance. The government's emphasis on the role of the GP and on primary care may be seen as an attempt partially to redress this balance. Such policy initiatives are unlikely to be challenged by the nursing profession, since they do not traditionally have the same power base or degree of influence over government policy. Most recent policy offers the prospect of limited access by nurses to the new commissioning groups, but in practice this may prove difficult to achieve. If it is the GP who explicitly or implicitly controls the definition of nursing roles then, as a profession, it becomes more difficult for nurses to define clear pathways and frameworks for professional development. This does not just relate to the future, but to the present as well, since the research suggests that many nurses in primary care have skills which are not utilised or are underutilised or that their skills and training may not necessarily be suited to the jobs they are required to do (Ashburner *et al.*, 1997). There are other dangers if nurses lose such control, especially in relation to the scope for innovation in nursing practice.

The nursing profession is currently promoting greater unity and cohesion across the various internal groupings. However one has to question whether any significant changes will be effected without the support of the medical profession. Within primary care, a high degree of control remains in the hands of the medical profession. In the acute sector, the balance of power has been changing as the level of managerialism has increased, but despite this there has been no major loss of medical power – merely a realignment and a greater involvement of clinicians in management (Ashburner, 1996). In primary care, however, the level of managerialist intervention is still limited. Until a similar balance is achieved between the influences of managerialism and professionalism, professional boundary issues, rather than a clear definition of health need and the most appropriate modes of delivery, will continue to influence the configuration of primary care services. There is no comprehensive and cohesive management structure within primary care. The extent to which the future influence of community trusts is able to expand is still open to question. The health authority, as the main purchaser, has set contracts with providers and this has been a key driver of change. Now that an increased commissioning role is being given to GPs there is potentially a reduction in the power of health authorities to influence the behaviour of GPs. Under such circumstances the future interests of the nursing profession can be seen to be potentially further weakened.

There is an urgent need for a shift of focus towards research based upon the needs of primary care and the role of nursing. The provisions of the 1997 white paper for the inclusion of nurses in the commissioning process assumes this as unproblematic. However, given the existing power base of GPs and the more 'hands-off' role of health authorities, it is difficult to see how the nursing profession, in its current state of division and unreadiness, can play a leading role in the provision of primary care health care which is most appropriate for meeting the health needs of the population. Unless this is addressed, the development of the new primary care groups, which should be in place by April 1999, will continue to be driven by the more narrowly defined 'medical model' and will fail to capitalise on the opportunities offered through the contribution of nurses within primary care. The value of multidisciplinary and interdisciplinary research is that it can integrate the strands of research findings in the various areas of professionalism, managerialism and service development. Unless the significance of the nursing role is taken fully into account at the level of the primary care group, delivery of the most effective primary health care could be compromised.

References

Abbott, A. (1988) *The System of Professions: An Essay on the Division of Expert Labour* (Chicago: University of Chicago Press).

Ashburner, L. (1996) 'The role of clinicians in the management of the NHS', in J. Leopold, I. Glover and M. Hughes (eds), *Beyond Reason?* (Aldershot: Avebury).
Ashburner, L., K. Birch, J. Latimer and E. Scrivens (1997) *Nurse Practitioners in Primary Care: The Extent of Research and Practice* (Keele: CHPM).
Atkin, K. and N. Lunt (1995) *Nurse in Practice* (York: SPRU).
Brown, A. and D. Grimes (1995) 'A meta-analysis of nurse practitioners and nurse midwives in primary care', *Nursing Research*, 44(6), 332–9.
Calman, M. (1993) *Hospital Doctors: Training for the Future* (Report of the Working Group on Specialist Medical Training) (London: HMSO).
Davies, C. (1980) *Rewriting Nursing History* (London: Croom Helm).
Department of Health (1990) *GP Contract* (London: HMSO).
Department of Health (1996) *Neighbourhood Nursing – A Focus for Care. A Report of the Community Nursing Review* (London: HMSO).
Department of Health (1997) *The New NHS – Modern and Dependable* (London: HMSO).
Dowling, S., R. Martin, P. Skidmore, L. Doyal, A. Cameron and S. Lloyd (1996) 'Nurses taking on junior doctors' work: a confusion of accountability', *British Medical Journal*, 312, 1211–14.
Elston, M.A. (1991) 'The politics of professional power: medicine in a changing health service', in J. Gabe, M. Calnan and M. Bury (eds), *The Sociology of the Health Service* (London: Routledge).
Fawcett-Hennessey, A. (1991) 'The British Scene', in J. Salvage (ed.), *Nurse Practitioners: working for change in primary health care nursing* (London: King's Fund).
Freidson, E. (1970) *Profession of Medicine: A Study of the Sociology of Applied Knowledge* (New York: Dodd Mead).
Gabe, J., M. Calnan and M. Bury (eds) (1991) *The Sociology of the Health Service* (London: Routledge).
Greenfield, S., B. Stilwell and M. Druvy (1987) 'Nurse practitioners: social and occupational characteristics', *Journal of the Royal College of General Practitioners*, 37 (August), 431–45.
Griffiths, R. (1983) *NHS Management Inquiry*, DA 83 38 (London: DHSS).
Harrison, S. and C. Pollitt (1994) *Controlling Health Professionals* (Buckingham: Open University Press).
Honigsbaum, F. (1985) 'Reconstruction of General Practice: Failure of Reform', *British Medical Journal*, 290, 904–6.
Hunter, P. and D. MacAuley (1996) 'Is this the next step for nursing?', *Practice Nurse*, 23 February, 174–6.
Iglehart, J.K. (1987) 'Problems facing the nursing profession', *New England Journal of Medicine*, 317, 646–51.
Jackson, C. (1995) 'Nurse Practitioners: Testing the boundaries', *Health Visitor*, 68(4), 135–6.
Jordan, S. (1994) 'Nurse Practitioners – learning from the US experience: a review of the literature', *Health and Social Care in the Community*, 2(3), 173–85.
Leopold, J., I. Glover and M. Hughes (eds) (1996) *Beyond Reason? The National Health Service and the Limits of Management* (Aldershot: Avebury).

McKee, M. and L. Lessof (1992) 'Nurse and doctor: whose task is it anyway?', in J. Robinson, A. Gray and R. Elkan (eds), *Policy Issues in Nursing* (Miton Keynes: Open University Press).

Mundinger, M. (1994) 'Advanced Nursing Practice – Good Medicine for Physicians (Sounding Board)', *New England Journal of Medicine*, 20 January.

Parsons, T. (1951) *The Social System* (London: Routledge).

Quinney, D. and M. Pearson (1996) *Different Worlds, Missed Opportunities: primary health care nursing in a North Western health district* (University of Liverpool: HCCRU).

Robinson, J., A. Gray and R. Elkan (eds) (1992) *Policy Issues in Nursing* (Buckingham: Open University Press).

Rowley, E. (1994) 'The Role of the Practice Nurse', in G. Hunt and P. Wainwright (eds), *Expanding the Role of the Nurse, The Scope of Professional Practice* (Oxford: Blackwell Scientific).

Safreit, B.J. (1992) 'Health care dollars and regulatory sense: the role of advanced practice nursing', *Yale Journal on Regulation*, 9, 417–87.

Sailsbury, S.J. and M.J. Tettersell (1988) 'Comparison of the work of a nurse practitioner with that of a general practitioner', *Journal of the Royal College of General Practitioners*, 38, 314–16.

Salter, B. and N. Snee (1997) 'Power Dressing', *Health Service Journal*, 13 February.

Salvage, J. (1985) *The Politics of Nursing* (London: Heinemann Nursing).

Salvage, J. (1988) 'Professionalisation – or struggle for survival? A consideration of current proposals for the reform of nursing in the UK', *Journal of Advanced Nursing*, 13(4), 515–19.

Salvage, J. (1990) 'The theory and practice of the "new nursing"', *Nursing Times*, Occasional Paper, 24 January, 86(4), 42–5.

7 Enabling Leaders to Change: interventions with established GP principals through a mid-career break scheme

Virginia Morley, Nicki Spiegal, Faruk Majid and Priscilla Laurence

THE PROBLEM

Medicine features ahead of many other professions in terms of stress. This has been well documented and is reported in the literature as contributing to a high incidence of depression, suicide, social isolation, alcoholism and drug abuse (Quill and Williamson, 1990, Johnson, 1991). However stress at work is a complex problem especially for GPs. Work in the 1980s found that the main sources of stress for GPs included risk management of patients, isolation, poor relationships with other doctors and disillusionment with the growing awareness of the change in the role of the GP (Braithwaite and Ross, 1988).

The role of general practice changed rapidly throughout the 1980s, with a greater focus on health promotion and primary care teamwork. It is widely acknowledged that the 1990 GP contract changes increased paperwork for GPs (Department of Health, 1989). The 1991 white paper *Working for Patients* which introduced changes such as fundholding further reduced the morale of GPs currently in post and led to a dramatic fall in applications both for vocational training, now new registrar posts and in applications for partnerships (Vaughan *et al.*, 1997). Where retention and recruitment had been more difficult before, such as in inner-city London, the situation was exacerbated. Changes in the NHS were occurring simultaneously with policy initiatives elsewhere in the public sector. The effect of these changes in such areas as education and housing were that people with multiple needs found general practice to be one of the few remaining accessible resources within the welfare system.

London Implementation Zone Educational Initiatives (LIZEI) schemes were a response to this convergence of complex problems. They aimed to support initiatives promoting the recruitment, retention and refreshment of London GPs. As part of the LIZEI, the Mid-Career Break Scheme (MCBS) was developed within the Department of General Practice and Primary Care at Guy's, King's and St Thomas's School of Medicine.

AIMS OF THE PROJECT

The MCBS aims to offer GP principals during the middle years of their career, at age 37–55, an opportunity to take part in a number of specifically designed education and learning opportunities. It seeks to encourage them to continue their professional development and so refresh and maintain their interest and enthusiasm for general practice and primary care in south London. The objectives for the MCBS are as follows:

- to offer GPs an opportunity to take part in the design and structure of the MCBS;
- to offer GPs mid-career review seminars to allow them time to reflect on their careers and plan for the future both individually and in a peer group;
- to offer GPs a range of different learning opportunities including:
    - 'take a break?' – a one day a week programme which brings together an 'action learning set' with time to develop a project for six months; GP assistant/research associate cover is provided for GPs taking time out;
    - 'time for a change?' – a two and half day a week programme which brings together an 'action learning set' with more time to develop a project; GP assistant/research associate cover is also provided for the practice where the GP is taking time out;
- to evaluate the impact of these various interventions on those GPs taking part.

PHILOSOPHY

From the outset it *was* intended to take on to the scheme GPs who reflected the broad range of established GP principals working locally. The *scheme organisers* were particularly keen to ensure that the programmes *attracted* GPs who were not already associated with the academic department. Nor was the scheme *intended* to attract those considered to be at either extreme of the spectrum of general practice performance, either where their networks and interests already supported their change and development or at the other end of the spectrum where interventions other than education would be more appropriate to enable change. The *organisers* were committed to the view that existing GP principals in post locally are commonly in leadership positions both within their practices, and in relation to other health services. The shift of emphasis within the health service towards primary care and general practice as central to commissioning health services further supports them as leaders within the National Health Service.

It was recognised that, in order to fulfil these roles adequately, GPs would need time and space to adapt, take on new ideas and change.

The *programme organisers* also believed that the future held a vision of continuing change for general practice and that many practitioners needed time to adjust to this as part of their developing career in general practice and primary care. Lastly, there was an *interest* in designing and offering a scheme which not only supported professional development for GPs in the inner city but would also offer scope for personal development which would underpin opportunities for further change and development in the future.

DESIGNING THE MCBS

Using a learner-centred approach, the MCBS has been developed and built around the expressed needs and preferred working styles of local GP principals. In spring 1996, randomly selected groups of GPs were invited to take part in three focus groups to consider how the development of the MCBS could best meet their needs. The perceived benefits for GPs in taking part included improving their existing skills, for example use of computers, and developing their management and clinical skills. They also expressed considerable interest in developing new clinical skills such as minor surgical techniques.

They were enthusiastic about taking time out to reflect on their work situation, given the perceived pressures of workload. It was recognised that the MCBS could be a valuable opportunity to meet colleagues and reduce isolation. It was seen as a vital means of peer support with a group of new colleagues from different practices with whom they could build a network of professional support.

The focus groups challenged GPs to consider what would be difficult about taking part in a career break. Most of the issues raised related to their fears and anxieties about leaving behind a job of work that still needed to be done, combined with their uncertainties about what the career break might mean and how much such a change might affect them. Significant concerns were raised about leaving work; in particular GPs were worried about being seen to place further demands on their partners and the effect on colleagues who might envy their taking a break from day-to-day practice.

MID-CAREER REVIEW SEMINARS

From the focus groups it was evident that many GPs would welcome a 'one off' opportunity to reflect on their career in general practice to date, share ideas and begin to think creatively about their options for future professional and personal development: in short, starting to create a personal development plan and identifying the action needed to make it happen. This was intended to open up some of the options available which they could then choose to work on either as part of the MCBS or by undertaking educational opportunities outside the specific programmes of the MCBS, linking them into wider LIZEI programmes.

Two mid-career review seminars were run during the summer of 1996, attracting 17 GPs. Two groups which developed through the seminars also continued to meet over the next year. These seminars were important building blocks in the MCBS and similar 'one off' introductory events were run again in 1998. These seminars offered individual GPs a shared sense of current concerns in general practice, in particular offering initial insights into areas that cause anxiety and stress for GPs, such as complaints procedures and the challenge of establishing good relationships with their local health authority. The range of both personal and professional issues raised confirmed the programme coordination commitment to the design of the intervention, which, although focused on professional development, did not limit the possibility of GPs working on personal development issues where they felt this to be appropriate.

'TAKE A BREAK' AND 'TIME FOR A CHANGE'

'Take a break' is a one day a week programme lasting six months, with a two-day residential seminar at the beginning. It is designed around an 'action learning set' as a central component enabling GPs to learn new approaches to problem solving in relation to their work setting. GPs also undertake a supervised project of their choice.

'Time for a change' is a more substantial intervention with an individual GP taking time out of practice for two and a half days a week and providing three days of GP assistant/research associate time for the practice. To date four 'take a break' and two 'time for a change' programmes have been run.

'A typical action learning set is a small learning group which meets periodically over a drawn out period of time, allowing space for action between meetings. At meetings each person in turn is offered space and time to consider what they might do next in the light of their experiences since the set last met' (Casey, 1996). This model of groupwork was chosen as it seemed to fit most appropriately GPs' expressed needs. It offers an approach which explicitly links work experiences and problem solving at work. It is also based on a peer group approach and has been successfully used with others in the welfare system, including probation officers, social workers, health service managers and hospital doctors. As well as focusing on work issues it offers an important link to personal development issues. This focus on work-related professional development issues, balanced by opportunities for personal development, was felt to be an important boundary to work on with GPs.

Our focus of attention then was with GPs as leaders of organisations in primary care. It was assumed that an end point for these GPs was 'to get better at managing' and that this meant changing behaviour, not simply understanding management or learning about management. This was also the rationale for choosing a model of small group change which was not specific to general practice and did not explicitly focus on the doctor/patient consultation. Action learning was a concept originally devised by Reg Revans in the 1970s. David Casey (1996) comments, 'In action learning, Reg Revans has given us a development tool of great precision; when a group of peers has common development needs, emphasis can come off action and onto learning.'

Action learning also seemed an appropriate tool as it builds on Revan's concept of 'comrades in adversity'. GPs often find themselves in this position and this was an opportunity to provide a place where they could be themselves and find a way to manage their own learning. Casey discusses the particular need to manage chief executive learning. Many parallels in his discussion were observed which appropriately reflects the experience of GPs. For example:

> And so, over many years, chief executives develop a belief in themselves. They learn to think of themselves as somehow different. People near them begin to flatter as they see opportunity for themselves in the chief executive's power and soon a chief executive can become cut off from any trustworthy feedback. That is a very dangerous position to be in. Some get pushed beyond the point where it is difficult to say 'I don't know', 'I'm afraid' and 'I need help' to the point of no return, when they begin to believe that they do know, they are not afraid and they do not need help. That is an even more dangerous position. (Casey, 1996, p.13)

However the parallels between the experience of chief executives and of GPs in terms of their positions as leaders of organisations, at the top of the pyramid, go further. Casey reflects on the need for this work to become incorporated into careers much earlier on and the experience of working with groups of GPs also endorses this view.

> The final surprise for me is just how many chief executives are unhappy people. They have sacrificed too much, neglected close relationships, made wrong choices on the way up. Too often their steely eyes fill with tears and their work-weary faces collapse into regret. They need help earlier in their careers, help to manage their own learning so that they can make the broader-based choices they will not regret when they become chief executives. Programmes like Action Learning for Chief Executives come too late for some. (Ibid., p.17)

Each of the action learning sets has a mix of men and women GPs aged 37–55 and from practices of different sizes across south London. This allows for a range of different perspectives. Access to the programme is through written application and interview. The problems addressed have included clinical, organisational, professional and personal issues, providing both practical and emotional support. Topics discussed have included a wide range of issues from practice management and primary care commissioning through to bereavement and different models of consulting patients. Participants have also shared issues about the kind of practitioners they hoped to become, the ideals they work for and the realities and disappointments of general practice as they experience them.

Reviewing the action that has resulted for different individuals on the first 'time for a change' programme, it is possible to see changes in terms of GPs' future career plans, including options for part-time working, renegotiating partnership arrangements, future education and training and discussions with outside agencies such as providers and the health authority for different working patterns.

GP participants to date have also undertaken a wide range of different and contrasting projects including the following:

- developing working relationships with other professionals;
- leadership in a general practice setting;
- locality commissioning and fundholding;
- developing group facilitator skills;
- researching the teaching of sexual health in junior and secondary schools;
- sports medicine;
- mental health issues in a deprived population;
- developing personal assertiveness.

These are chosen by the individual GP but shaped and developed through the input of the project supervisor.

DISCUSSION

The evaluation of the MCBS attempts to examine its structure, process, and outcome, thereby assessing its success in achieving its objectives. While data for all groups of GPs on both 'take a break' and 'time for a change' are being collected, data are only available for the first group of GPs who took part in 'take a break' at the present time. GPs taking part reported expecting the MCBS to provide protected time in which to pursue a project and to reconsider and reflect on their job. They also expressed a wish to renew their commitment to medicine. In addition, it was felt the MCBS might provide new incentives and broaden horizons.

By the end of the scheme, in almost all cases, there were positive feelings about having taken part. GP participants had been able to identify problems and discuss them, and most had realised change was possible. In some this had already been implemented by changing working practices, including moves to part-time work. The evaluation also sought to gather quantitative data and different aspects of the GPs' life and practice have been measured using questionnaires applied before and after the programme. These included the occupational stress indicator, the GP attitude questionnaire, the general well-being survey and an inner-city practice questionnaire.

At the start of the 'take a break' programme this group of GPs spoke of stress linked to their participation in decision making; flexibility in their working lives; the way in which they were able to deal with change; implementing policies; communication; and their personal relationships at work. The range of stress perceived was wide but most was experienced from a high workload, patient expectations and what were felt to be unreasonable demands from patients.

Analysis of changes in all questionnaires suggest that, although by the end of the 'take a break' programme perceptions of GPs taking part of general practice as a stressful job, the extent of the applied stress and ideas about the sources from which that stress came were unchanged, their skills for coping with potential stress, job satisfaction and physical health had all improved.

It is impossible at such an early stage, with only limited data available, to draw conclusions. However, a number of issues have arisen which warrant further discussion. Current policy supports general practitioners in a leadership role within and outside the practice. While there is growing support, particularly in the changes proposed under the 1997 white paper (Department of Health, 1997) for a wider sharing of leadership responsibilities across the primary care team, the current structure endorses GPs in a central role with regard to patient care in co-ordinating a range of primary care and secondary services. In recent years the greater emphasis on primary care within the NHS means there is an increasing expectation that GPs will influence the delivery of health care. This has been described as a desire for a 'superdoc': clinically competent, available and accessible to patients, organisationally and managerially efficient with a vision of the future for service provision, and able to communicate this to a wide variety of colleagues and the public. While there may be many GPs who are skilled enough to undertake this challenging and ever-changing role, there are also many who are not. Furthermore the position of general practice within the health care system as a network of small organisations means it is often difficult for individual GPs to succeed in this multifaceted role.

The current pace of change and level of continuing uncertainty in the health service – not only in the inner city – suggests that many professionals in leadership positions, such as dentists, pharmacists and hospital consultants as well as others in the primary care team, may benefit from these kinds of interventions designed to refresh and to maintain individual adaptability to continuing change.

It is also clear that some of the changes introduced at the beginning of the 1990s may also have reduced opportunities for peer support between practitioners across different practices and that this may have contributed to worsening morale and greater social isolation of individual practitioners at a time when they are under pressure to manage an increasingly complex, clinical workload. Reinstating structured opportunities for establishing networks between practitioners is clearly going to be essential if the setting up of primary care groups in any shape or form is possible.

For most practitioners on the MCBS programmes this is likely to be a once only opportunity to take part and many have already been disappointed as the programmes have been oversubscribed. We suggest that such opportunities would be most likely to be effective if they were offered as part of a continuing programme of professional development throughout a GP's career, starting when he/she enters practice and continuing at intervals throughout his/her career. As argued earlier in this chapter, for many GPs, as for chief executives, opportunities of this kind will have come too late. This is important and linked to concerns that GPs have often appeared to find it difficult to come forward. Applications for the programmes have increased considerably over time and some GPs expressed real concern that they would become stigmatised as 'unable to cope'.

It is also important to note the limitations of this type of intervention. So far it has not been possible to measure the impact of a partner leaving a practice and how partnerships and practices may have changed as a result of this. It may also be that the influence of having a younger GP (Harrison and Van Zwanenberg, 1998) replacing an older established partner has been important to some practices and it may be that such interventions should be incorporated as part of a whole range of support measures for practices. The recent report by the Chief Medical Officer (1998) offers a vision of an education structure which is practice-based, interdisciplinary and related closely to patient needs.

The Mid-Career Break Scheme has identified the importance of participation by GPs from different practices in enabling doctors to draw on wider perspectives. The question of the optimal balance between unidisciplinary and multidisciplinary learning is complex and unresolved. Whether the GPs who attended the MCBS would have achieved similar mutual trust in a more mixed group is an interesting question, worthy of research.

Finally the link between healthy doctors and their capacity to assist in the delivery of a health care system which meets patients needs should not be underestimated. The traditional structure of the general practice career has been shown to lead to 'burn out' in alarmingly high numbers. The Mid-Career Break Scheme may play a part in the prevention of this waste of talent.

Acknowledgements

The MCBS involves a team of individuals who contribute both to the dynamic nature of the scheme and also to its overall success and evaluation. We should therefore like to thank and acknowledge all the GPs who took part, the GP assistants/research associates and Annie Atherton, Christine Bell, Tim Dartington, Tony Emerson, Lesley Higgins, Roger Higgs, Monica Martin, Tricia Scott, John Seex, Alison Wertheimer and Patrick White.

References

Braithwaite, A. and A. Ross (1988) 'Satisfaction and Job Stress in General Practice', *Family Practice* 5, 83–93.

Casey, D. (1996) *Managing Learning in Organisations* (Buckingham: Open University Press).

Chief Medical Officer, Department of Health (1998) *A Review of Continuing Professional Development in General Practice* (London: HMSO).

Department of Health (1989) *General Practice in the National Health Service: the 1990 Contract* (London: HMSO).
Department of Health (1997) *The New NHS – Modern and Dependable* (London: HMSO).
Harrison, J. and T. Van Zwanenberg (1998) 'GP Tomorrow', *Radcliffe Medical Press*.
Johnson, W. (1991) 'Predisposition to emotional distress and psychiatric illness amongst doctors: The role of unconscious and experiential factors', *British Journal of Medical Psychology*, 64, 317–29.
Quill, T. and P. Williamson (1990) 'Healthy Approaches to Physician Stress', *Archive of Internal Medicine*, 150, 1857–61.
Vaughan, C., D. Tovey and N. Foxton (1996) 'GPs on work, rest and play', *Health Service Journal*, September, 24–5.
Vaughan, C., J. Hitchens, V. Morley and M. Carlisle (1997) 'General practice training in south London: the South London Organisation of Vocational Training Schemes', *Education in General Practice*, 17 August, 251–4.
Working for Patients (1991) (London: HMSO).

8 Medical Managers: puppetmasters or puppets? Sources of power and influence in clinical directorates

Lorna McKee, Gordon Marnoch and
Nicola Dinnie

INTRODUCTION

The theme of the relationship between the medical profession and management has been at the centre of both academic and policy debates since the creation of the National Health Service and affects the design of most health care organisations worldwide. It surfaces and gains prominence when any new structural arrangements are in prospect, both shaping and constraining the way problems are defined and tackled (Buchanan *et al.*, 1997). Usually there is some kind of accommodation between political interests, managerial or bureaucratic necessity and professional capabilities and demands. The overwhelming challenge is how to establish appropriate governance and how to integrate and organise 'those responsible for medicine with those responsible for money' (Dawson *et al.*, 1995, p.172); 'those spending and those managing money' (Strong and Robinson, 1988). There are also the related questions of who should manage 'experts', how to reconcile corporate and professional interests and whether health care organisations pose a different managerial task from other organisations.

According to many commentators, different structural solutions have encouraged one or the other constituency of manager or doctor to win in the contest for power and dominance (Friedson, 1970; Marnoch, 1996). Disagreements have emerged in interpreting whether managers or doctors have gained ascendancy at any given period. For example, at present there is the view that the general management changes of the 1980s and the market-based reforms of the 1990s in the UK succeeded in eroding medical power (Flynn, 1992; North, 1995). Alternatively there are those who conclude that doctors have subverted the current management reforms, rigging the management agenda to further their own interests and ensuring that their professional priorities have gone unchecked (Hunter, 1991; 1994). The role of the other 'clinical trades' is usually taken to be of secondary importance or at best subdued in this struggle (Strong and Robinson, 1988) and discussed in Chapter 12 of this book. In the USA there has been a growing literature on 'deprofessionalisation', suggesting a shrinking power base for doctors and other professionals (Haug, 1993) while in the UK some recent observers are posing the contrasting notion of 'demanagerialisation' of doctors, although it is a particular market–managerial tradition that is said to be in retreat (Ferlie, 1997).

A third strand is now emerging from a number of empirical studies which suggests that, while at a collective level it is useful to continue to think of doctors and managers as adversarial superpowers, the microlevel reality is more complex and reveals some fascinating compromises, alliances and innovations. Doctors and managers are looking less polarised, coming together in multidisciplinary non-hierarchical project teams, and adapting to external threats by spanning conventional boundaries. New organisational forms and processes are revealing a more comfortable blend of corporateness and professionalism (Fitzgerald and Dufour, 1997; Ong *et al.*, 1997). This prospect of doctors and managers being diverted into strategic liaisons, taking entrepreneurial action and creating new organisations has been especially evidenced both in the USA and in Canada (Marnoch, 1996; Fitzgerald and Dufour, 1997). However, Ong and her colleagues (Ong *et al.*, 1997) provide UK-based examples of the reconfiguration of cancer services and the concept of disease management which have resulted in cross-speciality and multiprofessional linkages. These findings are intriguing, leading to the speculation that doctors in the UK (especially in the post-market era) could be more directly drawn into crafting new organisational forms and pushing for radical organisational change (Marnoch, 1996). What is less clear is the contexts and actions that stimulate and nurture these unconventional developments.

In this chapter, the issue of the interface between clinical and managerial realities is addressed by analysing the managerial complex at the heart of clinical directorates. Clinical directorates were a device aimed at directly engaging doctors in the management process and have resulted in the creation of the combined role of 'doctor–manager'. They were the 1990s response to those issues outlined above concerning accountability, medical leadership and executive control. Clinical directorates were based on a model first pioneered at the Johns Hopkins Hospital in the USA. The model based on specialist or functional units with independent cost centres and devolved budgets was first adopted by Guy's Hospital in London. With minor modifications the model or a variant have now diffused rapidly and widely across the UK. The clinical directors are usually senior-level doctors who retain their clinical duties but gain responsibility for a unit of management. They are usually a part of a directorate management team which includes a nurse manager and a business manager referred to as the 'triumvirate' or 'triad' (Fitzgerald and Sturt, 1992; Hunter, 1992; Fitzgerald, 1994; Dawson et al., 1995; Dopson, 1995; Buchanan et al., 1997).

Drawing on a study of clinical directorates in Scotland, this chapter addresses the context, content and process of clinical management. Clinical directors are taken to be indicative of the mood and conduct of medical management in the UK and represent one constituency who can provide immediate insights on the wider conceptual and political challenges of organising and managing health care services. The chapter has three key sections. First, it compares and contrasts different clinical directorates and highlights organisational complexity. It is argued that, while clinical directorates appear to follow a structural blueprint and display an apparent organisational uniformity, distinctive directorate cultures evolve. These different directorate cultures might support the medical dominance thesis, indicate some growth of managerialism or reveal a degree of power sharing between doctors and managers and others ('the adaptive model', as described by Ong et al., 1997). The data indicate that team approaches to medical management and effectiveness may vary from trust to trust and even within trusts. This focus on differences and uniqueness contrasts with the many recent studies which have concentrated on similarities and overarching patterns.

Secondly, the chapter will pinpoint the multifaceted and often subtle influences on key organisational players and forms. The aim here is to tease out the sorts of conditions and contexts which determine directorate life cycles and cultures. Are there circumstances which break down or reinforce the conflict model of doctor versus manager? Using data from detailed case studies, it is argued that knowledge-based organisations such as the NHS can sustain competing realities and that a mix of circumstances, problems, personalities and biographies can generate multidimensional responses. The haste with which clinical directorates were established has meant that evaluations of their operations are only now becoming available.

In the third section, the chapter will not just reflect on the diversity and the conditional nature of emergent clinical managerial realities, but will add to the discussion about future forms of medical management and organisational development for the millennium (Marnoch, 1996; Fitzgerald and Dufour, 1997; Ong *et al.*, 1997; Ferlie, 1997; Hunter, 1997). It will analyse which ingredients appear to lead to effective directorate teams, indicate what trusts might do to exploit medical leadership and innovation, and highlight which conditions seem to foster meaningful collaborations and the successful management of change. The future role of the directorate business manager will also be reviewed, as it has often been overshadowed in earlier analyses of the clinical director, whom researchers have regarded as the primary actor in the directorate drama.

Methods

The chapter draws directly on an Economic and Social Research Council funded study of clinical directors in Scotland who were surveyed through a postal questionnaire (response rate 65 per cent) and detailed case studies of a sub-sample of directorate personnel. In particular it compares and contrasts different directorate management processes and power configurations across six NHS trusts in Scotland and draws on in-depth interviews with key personnel at different tiers in each organisation (in total, 50 interviews were achieved, with 24 clinical directors, 13 business managers, three nurse managers, four chief executives and six medical directors). The trusts were geographically dispersed and included community and acute trusts in urban and rural locales.

Categorising directorate management realities: what did they look like?

In developing a comparative analysis of the six trusts chosen for more in-depth study, one of the striking findings from the present study was the marked variability both in the way clinical directorates were constructed and in the way they conducted their business. Many empirical studies to date have rightly aggregated and emphasised the common experiences of clinical directors (Dawson *et al.*, 1995; Fitzgerald and Dufour, 1997) or have drawn their data from a single locale (Buchanan *et al.*, 1997; Willcocks, 1997; Thorne, 1997). Here it is possible to construct comparative cases which have been shown elsewhere to be a powerful tool in understanding contextual aspects of change and the management process (Pettigrew *et al.*, 1992). During the qualitative phase of the study, the observations and interviews showed that, behind fairly streamlined organisational charts, common objectives and often apparently identical directorate models, there lay subtle differences in interpretation of the task, contrasting internal and external environments, different managerial orientations, competing change agendas and a wide range of human competencies/expertise and actions. The pace of change, directorate 'energy' and proactivity levels varied markedly from place to place. The complexity of the directorate agenda and day-to-day business also revealed marked contrasts. The picture that emerges is one of negotiation and fluidity, which contrasts with the simplistic view of 'top-down' uniformity of directorate models. Clinical directorates will be shown to provide both change and continuity and to offer both opportunities and constraints.

TRADITIONALIST, MANAGERIALIST OR JOINT VENTURER?

In analysing the qualitative data, it was useful to assess whether directorates (a) displayed a bias to medical or managerial agendas, (b) created new organisational opportunities, (c) were involved in innovations in both process and services, or (d) planned strategically or were intent on maintaining traditional allegiances and hierarchies, 'ways of doing things' and priorities.

The context of the directorate and the nature of its business were also examined in an attempt to classify their orientation (Pettigrew *et al.*, 1992). From this work on the qualitative data, at least three major directorate trends emerged, although even here there is a danger of oversimplifying the reality, and the categories themselves should be perceived as 'ideal types' subject to anomalies and contradictions. The analytical separation is used primarily to highlight the intricacies involved in constructing 'doctor–manager' roles and in attempting to integrate professional and managerial domains.

Traditionalist

The most dominant and prevalent impression was of those directorates which were struggling to 'keep the show on the road'. As such they were largely preoccupied with external threats in the form of financial crises, cost constraint and the search for efficiency savings. There were also merger issues, downsizing, competition from other clinical groupings and the reconfiguring of services to confront (especially true in the acute sector and in the very large directorates). The scope for clinical directors to innovate or influence strategy was highly constrained in these circumstances, as one reports:

> I mean the biggest disappointment to me as a clinical director is to see that no matter how good or innovative you are, at the end of the day you are at the mercy of people who are counting pennies all the time... I must admit the cynical part of me thinks that talking about having visions or what, it is all very nice, but the reality is, there is no money.

The recurrent theme in these directorates was the enormity of the task and the paucity of management resources relative to expectations. These structural constraints resound in the words of one clinical director:

> I think one of the problems we directorate teams have and ours suffers from, is to know how big it should be. What staff do you actually need to run a staff of 750 and a 20 million budget? If you actually think it's one and a half battalions in military terms, if you were to think what the structure within a battalion is and how many people it takes to run a battalion we are kind of doing it on a bit of a shoe string.

These trusts could be described as 'traditionalist' in that they were low on both process and organisational innovation and the clinical director was primarily attached to a clinical/professional perspective. Even when there were high-calibre clinical directors in place, the sheer operational demands constrained their managerial ambitions. A number of clinical directors here did express a frustration with the job, found the dual demands excessive and had no long-term desire to remain in managerial positions.

> I certainly wouldn't leave medicine to shuffle paper for the rest of my days; If they were asking me to stop doing clinical work for 2 years to be a clinical director, I would say no because basically I might not be here in 2 years, I might be run over and it would be dreadful run over just going to meetings.

> Sometimes I think maybe I should just stay in management because going to committee meetings is less stressful than dealing with difficult clinical problems but I don't know if I could go to committee meetings for the rest of my life because it tends to be a bit tedious.

Traditional medical hierarchies and groupings tended to remain intact and clinicians retained a fair degree of control over priorities. Usually there was little time for reflection or strategic thinking and there was a culture of firefighting rather than opportunity seeking. One clinical director sums up this 'maintenance' role:

> Most people want someone to keep the show on the road and to respond when they give it a kick in a particular direction. It is actually easier to do the administrative jobs and we probably spend too much time on that and any management actually comes out of the inevitable rather than sitting down with a blank sheet of paper type thinking.

Such an environment seemed to both protect and preserve conventional structures, with other groups of clinicians only becoming loosely aligned to the directorate management team and task. In these directorates there was only minimal involvement, if any, of the clinical director in the contracting process, marketing or in external negotiations with other tiers of management, sectors or agencies. One clinical director summed up this distance from contracting thus:

> The whole thing is basically done by this chap who works with the Health Board and then they tell us what they've done, even if that is what we have told them they can't do.

This supports earlier findings by Dawson and her colleagues (Dawson et al., 1995), based on data gathered in 1992, and demonstrates a continuing inability or reluctance of clinical directors to engage in all aspects of the NHS internal market. One clinical director had had his 'knuckles rapped' for trying to take another trust's business away. There were certainly some grounds for asserting that the clinical directors here were 'directors' in name only, evidencing little impact on the trust's direction, traditional power bases or professional conventions. However, even in these stable low-change conditions, some clinical directors recognised a degree of change in themselves. One clinical director commented:

> I have got a bit more of an idea of the cost of things; I have learnt a bit more understanding about what is involved in management... because basically I have to confess you were never really bothered about how the machine ticked along, you just did your bit.

Another clinical director commented:

> I think I have become much more aware of financial issues and of the interrelationship between departments and their necessity to prioritise because we only have a certain amount of money.

The nature of the relationships with clinical colleagues in these directorates was often described as based on a consensus-building role or that of facilitator. The clinical director in these cases relied heavily on persuasion, 'sweeteners' and only very rarely on sanctions. The relationships were still embedded in a collegiate clinical network and not formed around corporate organisational interests per se. The emphasis on co-ordination and consensus comes through in the following comments:

> We are not actually called clinical directors because we cannot really direct what goes on... I don't say to people, this is what I want you to do or this is what I want to happen. I say to them tell me how we can best work it out? There is no way I am imposing my will.

The clinical directors preferred to negotiate with clinical colleagues rather than attempt to command, recognising constraints on their formal ability to hire and fire: 'At the end of the day they know I can't go and sack them.' While a number of clinical directors admitted that the main device they had at their disposal was the ability to authorise funding to colleagues, even this 'power of the purse strings' was constrained and related mostly to new activities at the margins of the directorate budget.

> Consultants see one another as equal and to an extent management decisions have to be made and I would suggest it to them and hopefully effect the change, but they could resist it strongly and you can't sack them if they don't do what they are told. You have got no power of authority over them really, you really have to take them along with decisions. And the bottom line tends to be funding, you say, well we haven't got the funding for that, so you become unpopular in that respect. It can be difficult.

While in 'traditionalist' directorates it was possible to find highly skilled and experienced business or support managers, they often reported high frustration levels and a sense of powerlessness. They were being expected to 'minesweep', create a climate for survival rather than develop new territories. The clinical director typically occupied the position at the apex of the management team triangle and the business manager and nurse manager were directly accountable to him/her. The 'knowledge or technical' apex was often felt to be inverted, with the business manager having greater yet undervalued expertise in areas such as appraisal, 'people skills', negotiating, budgeting and planning. Yet the clinical directors were seen as failing to deliver on the one area where they were supposed to make a difference, such as influencing and managing their clinical peers. These tensions were often deep-seated and resulted in business managers in traditionalist directorates querying the efficacy of the whole directorate model and philosophy:

> I am a minder for the clinical director and that is a word I have heard other business managers use. We have to feed them information, we do the work for them, we write papers for them, we come up with efficiency plans for them. We do all of that and they take the glory. I mean if it goes well they are the ones on the management group who get the credit for it and if it goes badly we are the fall guys and they can go back to their day job.

The disillusionment with clinical directorates is reinforced by another business manager:

> I have come to the conclusion that, as far as I have seen, it isn't adding the kind of value it should or resolving the problems or really attacking the resources used by clinicians, that would be the point of clinicians in management. I don't see any point in having a figurehead that is an expensive figurehead. If that is all it is, then it is a very expensive way of managing and maybe at the end of the day they are better being doctors, that is what they are trained to do, so maybe it is actually better value for money that they go and be doctors and treat patients.

Managerialist

The second category of clinical directorate affords a stark contrast with the above. Here there was an impression of general managerialist principles having been adopted and some sense of clinical directors being proactive, grasping opportunities beyond their local directorate constituency and professional base. By 'managerialist' it is meant that the directorate was overtly concerned with the issues of efficiency, control of professional work, new relationships – with service users and between professionals and managers – and the application of business techniques (Cohen *et al.*, 1997).

Only a few of the 24 directorate cases fitted this typology, although some managerialist indicators could be found in a larger number of cases. In one or two instances those fitting this classification were very large directorates, their budgetary clout perhaps bolstering their management legitimacy and emboldening their commercial sensibility. If they had been involved in takeovers or mergers with other specialities, a superpower bloc was often created. This could bring enhanced exposure to the trust board, other directorates, purchasers and commercial realities. One clinical director explained that, rather than passively interpreting clinical directorates as an unwelcome extension of managerial control to the domain of doctoring, the introduction of clinical directorates should be recast as an attempt to empower clinicians. Doctors should properly be engaged in management in this version and delivering services and managing them are perceived as inseparable.

The reasoning behind the setting up of clinical directorates is to make clear that this is a clinically-led organisation. The people who are central to the delivery of the service ought also to be the people who are central to the management of the service.

'Managerialist' directorates typically had strong links to top management; the clinical director had direct access to the chief executive and a power base external to the clinical domain. This convergence of medical and managerial domains could draw them closer into broader organisational considerations such as strategy formation or contracting; in short, raise their managerial consciousness. Kitchener, in his analysis of two large Welsh trusts, also found examples of clinical directors who had 'begun to lead contract negotiations' (Kitchener, 1997, p.17). Size did seem to be a factor in the likelihood of a directorate gaining a degree of managerial prowess and in one instance, a medical director noted paradoxically, the danger of enhanced directorate autonomy and devolution, leading eventually to isolationism from broader trust strategy or other directorates: 'you have got larger entities that don't talk to each other instead of smaller entities'. This was a case of greater management control at a local level leading to deeper tribalism between specialisms.

The roles of the chief executive and medical director, the support and executive structures created for clinical directors, and communication patterns were also influential. Certain trusts seemed to be both more inclusive and positive about the potential of clinical directorates and tried to create a responsive infrastructure to realise this potential. One chief executive described his awareness of the clinical directors' limited power in controlling the behaviour/spending of consultant colleagues and outlined the steps he was taking to address the issue:

> The difficulty that clinical directors have especially in the large directorates of medicine and surgery is effecting control over clinical practice within individual specialities ... There is no line so it is difficult for the clinical director to exert control over the level of spending that happens within the individual services. We are reviewing that at the moment and we have a draft revised job description for heads of service, with the intention of pulling them into the management structure.

This chief executive planned to appoint heads of service 'rather than have them elected on a Buggin's turn basis'. The 'managerialist' clinical directors often displayed enthusiasm for their jobs, could be quite 'hands on', were fascinated by organisational dynamics, and often had considerable professional standing (one was described as the most senior surgeon in his hospital), having possibly held head of service or other senior management posts in the past. They had acquired a 'taste' for management and would perhaps have occupied leadership positions in other contexts. They had typically strong lines of communication with senior management and other clinicians. In exceptional cases, they aspired to a long-term managerial career. These clinical directors reported a sense of achievement in their jobs. One suggested that, owing to his influence, previous medical advisory structures had been integrated and superseded by the directorate structure: 'in the surgical directorate the advice and the management is now through me, through the heads of service in the directorate'. He expressed his high level of commitment to the role as follows:

> I play a major hands-on job here. I do quite a lot of the administration things. I work very long hours in fact... I mean there are ups and downs. It is quite interesting. I mean one does it because you know what is going on. I like to know, to have my nose into things. I like to see things developing but I am also aware that in terms of time commitment it is ridiculous. In terms of family commitment, my wife is extremely accommodating.

Another defined his job as 'dynamic', while a third felt he did not 'necessarily see myself going back into full-time clinical medicine' and that he might be persuaded to stay in management 'provided I haven't had a stroke, a coronary, ulcers haven't burst'.

Curiously, even in the more managerialist contexts, there was considerable organisational continuity and a failure to experiment with new organisational processes or forms. While clinical directors were seized by the managerial agenda, the climate was generally conservative. Their aspirations were to do with drawing bigger boundaries rather than spanning them. It did not involve 'smashing or remaking contexts' or working with new constituencies, although the traditional interface between medicine and management was challenged. As with the 'traditionalists' the focus of the directorate energies was parochial, internal or, at best, 'within trust'. There were also important checks on how far managerialist interests could go. Heads of service, university departments and teams, parallel medical political power groups of clinicians (for example, area committees of physicians) could constrain even the most 'bullish' of clinical directors.

Business managers could be very robust and strong influencers in managerialist directorates and described themselves as 'leading from behind'. On the other hand, and rather ironically, they sometimes described being isolated in this context, having felt their domains to be hijacked, bypassed and their own functions compromised. Some chief executives, in an attempt to 'capture' their clinical directorate constituency, created management reporting lines or an 'open door' approach which, perhaps unwittingly, seemed to marginalise business managers. Such tensions in directorate management could undermine effectiveness. One business manager reported that some consultants continued to bypass the directorate and that the chief executive weakened the devolved management arrangements by responding directly to their wishes: 'the damage to some extent has been done and will take a long time to undo because people know that he authorises things'.

Power-sharing or joint venture

The third category of directorate came close to a power-sharing model. Although in its pure form none of the case study directorates could be said to be represented in this way, a few directorates did reveal innovative ways of working and initiatives which crossed traditional speciality lines, flattened hierarchies and broke the mould of the adversarial medical–managerial stereotypes. These directorates were often focused on developing new services or reconfiguring services and, in one instance, previously warring groups of specialities were joining to provide a new integrated cancer service. (This mirrors the observations of Ong and her colleagues.) One business manager describes the unifying and positive effects of such service innovations:

> We have some people who are excellent and who work well together. I mean the three specialities that are coming together in the BAN unit (cancer services) are working together very constructively. Of course they have got a huge carrot at the end, so they would see the directorate as being very helpful and facilitative.

However the fragility of these creative alliances was also underscored. There was a sense that rebellions could threaten at any time within even the most harmonious directorates, led perhaps by maverick consultants who retained the power of veto at all times. In one of the directorates where there was an aspiration towards a power-sharing model, the clinical director saw the role as more than a relabelling and as that of an evolving hybrid clinical–manager specialist. He wanted to influence and manage clinical colleagues, and to outstrip the commercially driven managerial agenda with its focus on performance indicators, target setting and commercial overtones. His motivation was to 'manage for the patient'. He was described by others as a gifted clinician and a gifted manager. There was a genuine will to recast the business of the doctor–manager and to create a new organisational environment with clinicians leading and delivering a 'managed service' in partnership with management. In this large complex directorate (with many separate interest groups) he referred to making the most of the multiple interdependencies and highlighted the value of criss-crossing lines of accountability and mutuality.

In another small, more unified, directorate, the clinical director also picked up the theme of power sharing and interdependence in the directorate triad. Referring to his nurse and business manager, he commented:

> They knew what the ropes were and they helped me out to start with. It's meant to be a line relationship but I see we are all equally able to make decisions. I think they tend to see me as making the final decision although it is not me that makes the final decision, often it's by discussion. If I strongly objected to something they would take heed of that but we tend to think pretty similarly.

The striking feature of power-sharing clinical directorates was the deliberate attempt to establish a team approach to problems and decision making, especially to integrate the 'top team'. This was sometimes more difficult to achieve than to aspire to and the complexity of democratising the decision process in the face of historical and structural inequality is evidenced in the account of the business manager (from the same directorate cited above):

> I have no true control as far as the direction of the clinical service. I feel the best I can do is prompt, push, shove, pull into what I think is the general direction... I don't have the clinical knowledge of where xxx services should go... It's more a kind of pulling the strings to get other people to do what you want them to do.

Another business manager in a large complex directorate describes the subtlety with which these teams negotiate their roles and responsibilities:

> He [clinical director] has a great understanding of how things should be managed. He's read Harvey-Jones and all the rest of it and he knows how things should happen. Nine times out of ten he knows the answer... he's a hands on person, he likes to do things himself. What I try to do with him is try to be his headlamp, because often in the NHS you are in a tunnel trying to see where you want to go and how you should get there. I am his headlamp. He has the visions, he knows what is needed and he understands the issues fully and he has the respect of everyone.

In the directorates with power-sharing characteristics, attention was paid to creating an organisational vision which was wider than the local directorate concerns. There was an awareness of the wider organisational setting and an aspiration to break down boundaries, both professional and organisational. There was a climate of proactivity. For example, in one directorate, a study day to explore directorate strategy was organised as a deliberate attempt to help to push forward the directorate frontiers. It included both national and international clinical experts, management academics, the trust chairman and senior management, the dean of the medical school and other university representatives, other clinical directorates, local general practitioners, pharmaceutical representatives and the local voluntary sector.

Communication between the key members of the management team was reported to be good in these directorates and often involved regular briefings and both formal and informal contact. The clinical director seemed to have considerable 'political' awareness and was often partnered by a sophisticated business manager. This doubling of talent and ambition could be both effective and forceful in achieving organisational change. As with the 'managerialist' clinical directors, the connections with senior managers were also well developed in these few examples, and both formal and informal links were exploited.

In one of the case study directorates, the directorate team also favoured an 'open', 'up front' management style, and doctors and managers mixed both formally and informally. The business manager felt that barriers had been broken down and new alliances formed:

> Clinicians don't like secrecy because there is a mentality pervading through the service that management have always got something to

hide, something up their sleeve, there is some hidden agenda. So I try to be as open as I can... I am on first name terms with the consultants. Six or seven years ago you wouldn't get consultants socialising with people like me. I was dragged up to the pub there the other day just to be ridiculed at the bar; but there is change, don't get me wrong, there is a change and there is almost a camaraderie.

It was not possible to identify any directorates who extended this power-sharing tendency outside the framework of their own organisations. There were few examples of clinical directors who wanted to challenge the boundaries between primary and secondary care or between health and social care. Proposed mergers or collaborations of specialities were quite tightly delineated within a 'medicalised' context. It is hard to establish why even these more innovative directorates have failed to break down established hierarchies or to cross traditional organisational lines. It is possible to speculate that it is a reflection of the lack of maturity in the medical managerial initiative; equally it may be related to the coupling of the medical managerial initiative with the introduction of narrow competitive market principles.

WHY THE DIFFERENCES AND WHAT MADE A DIFFERENCE?

In this section we reflect on precisely what led to variations and continuities in directorates. What factors were involved in the shaping of the directorate reality? Who were the key players at a local level? Why did some directorates seem more flexible, innovative, proactive and 'successful' than others? Why was there evidence of more depth of change in some contexts than others (Pettigrew *et al.*, 1992)? What was the role of context and organisational structures and how far did individual action influence directorate outcomes?

The dimensions which emerge as important and meaningful points of comparison include, first, the role of key people both within and outwith the directorate. It is interesting to note that trusts were given very little central guidance on the human resource implications of directorates and largely made up their own guidelines concerning appointment systems, job descriptions, terms and conditions. Indeed the second wave of appointments indicated that some trusts were moving to a more formalised model of recruitment, selection and employment after a more ad hoc process. Of extreme importance was the background and recruitment circumstances of the clinical director. Here the issue surfaced of whether the post-holder had prior management experience (this did not necessarily mean in a clinical directorate post, but could be as leader of a speciality team, for example) as this could indicate a managerial orientation or flair. Equally important, as has been highlighted in other studies, was whether the clinical director was mandated by his/her colleagues or was a last resort choice: a willing or reluctant recruit to management. Further intangibles, such as: the clinical director's clinical standing amongst peers, seniority, membership of key political–medical networks, technical/management know-how, career stage and aspirations, were all found to influence directorate performance, culture and style.

The other critical internal post-holders shaping and affecting the directorate team were the business manager and nurse manager. Most of the trusts studied followed the triumvirate model where the clinical director was formally 'in charge' of staff and budgets but did not have direct responsibility for medical colleagues. Most frequent and extensive contact was reported between clinical directors and business managers. Yet there was again huge diversity in these posts and their incumbents. Significant differences occurred between those directorates which appointed full-time business managers and those where business managers were shared with other directorates or were part-time. The business manager's level of management experience and background was highly variable; some were raw recruits to the job, perhaps management trainees in their first assignment; others had worked for other directorates and brought cross-organisational skills; yet others were drawn from clinical backgrounds, nursing management or the professions allied to medicine. The status of the business manager showed just as many discontinuities as did their backgrounds and their aspirations. There were those whose next career move was to a director-level post and ultimately a chief executive position, while others were at the beginning of their NHS career trajectory. Lengths of service also varied and there were cases where the business manager 'turnover' was high, creating difficulties in management continuity.

Nurse managers displayed similar contrasts. Although fewer data were directly obtained here, the accounts indicate considerable variability. The impression was gained that the nurse manager role was more tightly circumscribed than that of the other two directorate team members, with clearer domains of operation and expectation. The traditional nurse–doctor relationship tended in many instances to freeze the nurse manager–clinical director relationships along predictable and hierarchical lines (Strong and Robinson, 1990). A number of interviewees denigrated the role of nurse manager, suggesting that, while the nurse manager had the biggest human resource management task, he/she rarely fulfilled a strategic role and came below the business manager in terms of power and influence. This view was not uncontested, but even some nurse managers felt they had slipped from the frame.

The chief executives (and to a lesser extent medical directors) could also be influential both in how they supported the introduction of directorates and in their preparedness to create new management structures which drew clinical directors into the broader organisation. The provision of clinical management review mechanisms, staff development/ training initiatives and dedicated recruitment/ appointment procedures were all indicators of how seriously the chief executive viewed the operation of directorates and his/her commitment to making them work. The trust infrastructure and climate could be manipulated in part to favour the effective operation of clinical directors. In one trust, for example, the chief executive had created a network of clinical directors with a direct and ready line to the senior management team (the term 'kitchen cabinet' was used informally to describe this newly created cadre).

The most powerful context for change was when a plurality of these key people combined: for example, a high-status, effective clinical manager with a talented and experienced business and nurse manager and a proactive chief executive.

Secondly, the history of clinician–clinician and clinician–manager relationships seemed to affect how far clinical directors could deliver change. There were many examples where the clinical directors appointed were neither senior nor well established. A chief executive lamented this fact: 'The calibre of individuals is very good but there is a problem in that the natural leaders of the service have not felt inclined to become clinical directors.' This could lead to existing clinical power blocs or hierarchies maintaining control and to parallel systems of decision making continually having the potential to oppose or undermine the directorate (Marnoch *et al.*, 1997). Equally, grafting clinical directorates onto a very conflict-riven medical–managerial context could be a tortuous business and act as a brake on the adoption of change (McKee,

1988; Pettigrew *et al.*, 1992). The converse was equally interesting and, where there were empowered, senior clinicians who could think managerially and strategically, the clinical directorate could be a force for change.

The role of strategy, shared values and organisational culture were associated with the way directorates conducted their affairs. In many trusts, clinical directors bemoaned the absence of any trustwide strategy and felt that the organisation was dominated by crisis and environmental downward pressure. These clinical directors felt disabled by the continual need to react to emergencies and were locked into old priorities and processes through the failure to plan. The term 'culture' is used to reflect the deeper assumptions, beliefs, ideologies, values and conventions which distinguished trusts (Pettigrew *et al.*, 1992) and in the NHS is not a single entity but is recognised as comprising multiple subcultures. It should be noted that in many of the trusts there was considerable resistance and suspicion toward, the market reforms, especially among clinical staff. Clinical directorates were often viewed as a symbol of this undesirable market-driven ethos. Importantly a number of clinical directors were reluctant recruits to the job, doing it often for negative reasons, either to oppose some other undesirable candidate or to stop a takeover by 'non-medical staff'. Furthermore what was striking from the data was the determination of most clinical directors to retain their clinical credibility, networks and means of return to a full-time clinical career. The power of these professional ties and the typically collegiate mode of relating to other clinicians were of strong cultural relevance, influencing how far clinical directors could go.

What is interesting is those few cases where directorates or trusts were forging a strategic approach. Here the role of individual visionaries, strong exceptional individuals and the personality of clinical directors (or others) did feature. The role of action and behaviour should not be underestimated in both the shaping of strategy and the subtle transformation of local cultures. The positional authority of the clinical director vis-à-vis clinical colleagues, that is the professional respect he/she attracted, could permit a degree of unconventional behaviour and risk taking.

The scale and nature of the business of the directorate and the clarity of its goals and tasks also affected the extent to which the directorate could innovate. The relationship here was not simple, uniform or predictable. Some of the very large directorates were consumed by financial crises which seemed to limit service or process change. These directorates often contained 12 or more specialities, each with its own head of service. Traditional priorities and modes of operating often persisted in the face of such complexity and there was the impression of the directorate structure as a close-knit system with only indirect influence. Yet again, there were instances of clinical directors wresting control of these complex contexts, focusing speciality interests, rejigging services and bed use, and cutting across normal advisory and accountability lines. The role of individual leadership and a supportive broader organisation stood out in these examples. If the focus of the change could be narrowed down, as with the integration of cancer services, this could also promote a clear directorate orientation. Other writers on the implementation of change have also drawn attention to the positive effects of 'goal consensus', the simplicity of goals and breaking change down into manageable and actionable pieces (Dufour, 1991; Grindle, 1980).

The final factor which seemed to affect the way clinical directorates evolved was a more abstract and conceptual one, relating to the way professional autonomy and managerialism were constructed and articulated by individual clinical directors (Cohen *et al.*, 1997). In short, the interpretive schema by which clinical directors moulded their roles was influential. The data revealed that clinical directors had internalised different versions of what the post meant and entailed and carried around different constructions and meanings of the concept of management and of professional medicine. For some clinical directors, management was perceived as an exotic set of technical skills and competencies which they often had yet to master and would acquire if only they had time, the right training and the aptitude. Management of budgets was usually perceived as a black hole that had to be quickly filled; administration was typically also cited as a core part of the skill set. The softer, more qualitative aspects of management – leadership, motivating others, personnel issues, political or negotiating skills – were less frequently raised or acknowledged. For these clinical directors there was a splitting of their identities, with medicine and management representing opposing categories. Few clinical directors admitted to having lost their dominant image of themselves as a doctor. There were also reported tensions between the 'science' of medicine and its exacting methodologies and the perceived vagueness of management and its tendency to faddism and rhetoric (Llewellyn, 1997).

Yet there were a minority of clinical directors who had abandoned this dichotomy and for whom there was a degree of ease and comfort in being a 'medical manager'. These clinical directors perceived the continuities in their roles of doctoring and managing and highlighted similarities in the roles and skills. They pointed to the way the experiences of being a clinician could be integrated into management contexts and to the transferability of expertise and knowledge. Although it is difficult to quantify this issue of the salience of the managerial identity to clinical directors, its impact should not be underestimated. Their versions of management and the deeper beliefs they hold about professional autonomy all surface when we evaluate why some directorates seemed more innovative, or more enduring, than others.

These perceptions and constructs of what management is or is not do not arise in a vacuum, nor are they fixed or immutable. Cohen and her colleagues (Cohen *et al.*, 1997), in a study of research scientists, similarly found a spectrum of adaptability and fluidity in their response to increased managerialism. The research scientists reported mixed definitions, ranging from managerialism as threat to managerialism as opportunity. Elsewhere writers have noted the skill of effective leaders in turning negatives into positives, problems into opportunities and panics into challenges (Pettigrew *et al.*, 1992).

DISCUSSION

The detailed findings about diversity and complexity in the clinical directorate 'experiment' are valuable for what they reveal about the way professionals perceive and respond to increased managerial control. They show that professionals are not passive in the face of 'top-down' change, that they can adapt, they can manipulate and reinterpret management diktat. The clinical directors in the present study could pull the strings as well as have their strings pulled, simultaneously puppets and puppetmasters.

The overall picture that emerges is one where some doctors are developing a degree of managerial sophistication, where their consciousness has been raised and managerial aptitudes have been learnt or enhanced. The 'top-down' initiative of clinical directorates has been recrafted at a local level and, through a blend of local action, creative behaviours and imaginative local structures, diversity and change have ensued.

Yet, despite these contextual differences and contrasts, there is also an overall sense of enduring structures and processes, of continuity rather than change. The traditionalist directorates were the most numerous and the power-sharing and managerialist directorates the most tenuous. The latter group could largely be said to represent exceptional individuals, backed by other constituencies of exceptional individuals, together creating and manoeuvring their change agendas and contexts. This is not to decry the importance of these circumstantial innovations, but it does raise the question of how well embedded these changes were, and how stable. Respondents themselves drew attention to the fact that the departure of even one key individual could threaten the integrity and functioning of a directorate. There was a sense that the attachment to the medical management model was weak, most doctors retaining a 'bolt hole' in clinical medicine and having made only a short-term commitment to the post of clinical director. The definitions of management were also fairly limited, with little strategic progress being advanced beyond trust boundaries.

There was little evidence in the study that the NHS as a whole had come to grips with the absorption of doctors into traditional line management/global resource control systems (Marnoch et al., 1996). Very few trusts had tackled the issue of short-term clinical director appointments, or of succession. There were examples of organisational 'lost learning' as very capable and inspiring clinical directors 'disappeared' into full-time clinical duties:

> There is short-termism. One of my big disappointments is that Hamish, the previous clinical director, has just melted away and gone back to being just another consultant. He hasn't continued to stick his head above the parapet to support management and to bring sensible management to bear. He has really just gone right back and I think he has just said, 'Thank goodness I don't have to do that again.'

This continued failure to embed clinical director innovation into wider systems was widespread. There were few examples of trusts creating a new climate in which clinical directors of the future were being spotted, nurtured or sustained. The drive was coming neither from management processes nor from within clinical contexts, although one should not diminish the efforts or long-term impact of external organisations such as the British Association of Medical Managers, the training initiatives of postgraduate colleges and individual trust in-house initiatives, influences on medical curricula and the existence of medical–managerial journals. Self-regulation through audit, evidence-based practice, clinical guidelines and numerous other 'self-management' initiatives also exert a transforming effect on clinical behaviour which should not be underestimated.

The overall context of financial constraint, crises and environmental uncertainty seemed to dog the introduction of clinical management. When run alongside the introduction of the quasi-market reforms, the introduction of contracting, fundholding and numerous other first-order changes it is easy to understand while so little concerted management attention has been directed at making directorates work or at understanding their dynamics. This is not to deny the noble efforts of a few visionary senior managers but to suggest that, in general, more attention was paid to clinical management design than to implementation or evaluation of its impact. As a major process change, clinical management has also been very thinly resourced, with many directorates run on a shoestring. The calibre of management support has also been equivocal. The roles of business manager and nurse manager and the construction of the 'top team' seem woefully in need of overhaul. There were managers who felt undervalued and deskilled, and the lines of support to finance and human resource managers were often dysfunctional. The management cadre themselves reported a degree of fragmentation and uncertainty. Again issues of career progression, succession and coaching for the future were neglected.

The post-market era may well release new opportunities for medical management to flourish and for attention to be paid to medical managerial capacity. For example, the reduction of the number of trusts and directorates might focus expertise and concentrate power in fewer hands. It is difficult to imagine that doctors will ever again be able collectively to ignore how money is spent or controlled or retreat to islands of managerial immunity. The clinical directors in this study had become 'socialised' into the concerns and perspectives of management, some more than others. Many managers also talked about finding it easier to relate to and appreciate the clinicians' world view. These 'elites' could be exploited. This convergence of the worlds of medicine and management and of empowered clinicians need not dissipate and could be reinvigorated and resourced. It will require deliberate actions, strategies and recognition.

What is less clear is how far clinical directors, or doctors in general, will become embroiled in the nitty gritty of creating new organisational forms and new networks, determining global budgets or radically challenging the overall shape of services (Marnoch, 1996). It would seem from the data in this study that clinicians could be developed into inspirational leaders and become active in forging new alliances. However more, and more senior, doctors will have to be given the incentive to get involved, the relevance of management will have to be actively marketed and the clinical legitimacy of doctor-managers will have to be safeguarded. Some policy direction, leadership and financing of this will be required. There will need to be a marrying of grassroot innovations with more central objectives. The commercial model of management and the language of private sector business will remain a barrier and management may have to be repackaged to appeal to professional and service ethics. In short a 'new medical management' will have to be created and the clinical directorates of the present transformed and reformed. Those managers who have developed a finesse in negotiating with clinicians and in building networks will also have to be picked out and encouraged. Such relationships and skills take time to mature and the time scale would have to carefully and realistically established. It would seem that clinical directorates are an episode in the change trajectory, influential but part of a climate-building phase. There are further acts to come as the NHS reaches a new post-market period with new policy imperatives.

Clinical directors and the organisational challenge

The NHS white paper has turned the role of the clinical director into a major policy issue. In what may be interpreted as a response to the problem of declining public and professional trust in the organisation of health care in the UK, the 1997 white paper has restated the 'doctors into management' policy using the new term of 'clinical governance' to describe a set of organisational arrangements and practices which will be required of NHS trusts:

> The Government will amend Trusts' statutory duties to make explicit their responsibility for quality of care. This will need to be taken every bit as seriously as the existing financial responsibilities... The intention is to build on existing patterns of professional self-regulation and

corporate governance principles, but offer a framework for extending this more systematically into the local clinical community, and ensure the internal 'clinical governance' of the Trust. (Scottish Office, 1997)

In short, chief executives are to be made accountable for clinical quality in the same way that they are accountable for financial performance and probity. Clinical governance will, according to the British Association of Medical Managers (the NHS doctor-managers' 'trade union'), require chief executives to develop the same type of close relationship with doctor–managers as they tend to have with their directors of finance (BAMM, 1998). This new set of reporting requirements, along with emergent relationships focused on delivering high-quality clinical services, can create a new clinical director-centred organisational discourse, based on an organisationally integrated and interdependent appreciation of health care (Marnoch and Ross, 1998).

In forecasting the organisational role to be established by clinical director in the post-1997 NHS, certain conclusions can be drawn from the preceding analysis. It should be remembered that clinical directors were initially seen as a means of giving chief executives some measure of control over activities carried out by doctors. Tightening hierarchical control was an issue of high concern, given the need to enter into contracts for the provision of clinical services to purchasers in the internal market. As well-paid, highly-trained doctors, clinical directors have provided poor value as 'hierarchy-binding' middle managers, a situation explicable in terms of their typical lack of status within local medical communities. Writers on organisational behaviour have been warning of the imminent death of middle management for some time (Keuning, 1998, p.24). The classic tasks associated with middle managers – the timely and accurate execution of operational activities – are made difficult, or in some cases impossible, by the complex power relations evident in NHS trusts. Few clinical directors attempted to fulfil a middle management role, hence the crafting of different roles described in this chapter.

As a discipline, organisational behaviour is being forced to move away from frameworks of understanding for which hierarchical templates are the basic building blocks. In the next century the formal job description of the clinical director may less resemble a poor impersonation of middle management and start to reflect the innovative practice evidenced in the behaviour of a significant number of the individuals studied in the research programme. Of key importance, given the emphasis on clinical quality contained in the 1997 white paper, may be the ability of the clinical director to effect reciprocal understandings of the clinical process between managers, patients and doctors. As important actors in the developing processes of clinical governance they have a role to play in stripping away the mystique surrounding the quality of clinical care and rebuilding the legitimacy of the NHS in the eyes of patients.

References

BAMM (1998) *BAMM Newsletter*, March.

Buchanan, D., S. Jordan, D. Preston and A. Smith (1997) 'Doctor in the Process... the engagement of clinical directors in hospital management', *Journal of Management in Medicine*, 11(3), 132–57.

Cohen, L., J. Duberley and J. McAuley (1997) 'Fuelling Discovery or Monitoring Productivity: research scientists' changing perceptions of management', paper presented to British Academy of Management Annual Conference, Queen Elizabeth II Conference Centre, September.

Dawson, S., V. Mole, D. Winstanley and J. Sherval (1995) 'Management, Competition and Professional Practice: medicine and the marketplace', *British Journal of Management*, 6(3), 169–81.

Dopson, S. (1995) 'Doctors in Management: a challenge to established debates', in J. Leopold, I. Glover and M. Hughes (eds), *Beyond Reason? The National Health Service and the limits of management* (Aldershot: Avebury).

Dufour, Y. (1991) 'The Implementation of General Practitioner Maternity Unit Proposals in Hospitals', PhD thesis, Centre for Corporate Strategy and Change, University of Warwick.

Ferlie, E. (1997) 'Demanagerialisation? An alternative future for the NHS', conference paper presented at *Medical Management: towards the millennium*, Department of Management Studies, University of Aberdeen, June.

Fitzgerald, L. (1994) 'Moving Clinicians into Management: a professional challenge or threat?', *Journal of Management in Medicine*, 8(6), 32–44.

Fitzgerald, L. and Y. Dufour (1997) 'Clinical Management as Boundary Management: a comparative analysis of Canadian and UK health care institutions', *International Journal of Public Sector Management*, 10(1/2), 5–20.

Fitzgerald, L. and J. Sturt (1992) 'Clinicians into Management: on the change agenda or not?', *Health Services Management Research*, 5(2), 137–46.

Flynn, R. (1992) *Structures of Control in Health Management* (London: Routledge).

Friedson, E. (1970) *Professional Dominance: the social structure of medical care* (New York: Atherton Press).

Grindle, M.S. (1980) 'Policy Content and Context in Implication', in M.S. Grindle (ed.), *Politics and Policy Implementation in the Third World* (Princeton, NJ: Princeton University Press).

Haug, M. (1993) 'De-professionalisation: an alternative hypothesis for the future', *Sociological Review Monograph*, 20, 195–211.

Hunter, D.J. (1991) 'Managing Medicine: a response to the "Crisis"', *Social Science and Medicine*, 32(4), 441–9.

Hunter, D.J. (1992) 'Doctors as Managers: poachers turned gamekeepers?', *Social Science and Medicine*, 35(4), 557–66.

Hunter, D.J. (1994) 'From Tribalism to Corporatism: the managerial challenge to medical dominance', in J. Gabe, D. Kelleher and G. Williams (eds), *Challenging Medicine* (London: Routledge).

Hunter, D.J. (1997) 'Medical Managers and Health Care Policy', conference paper presented at *Medical Management: towards the millennium*, Department of Management Studies, University of Aberdeen, June.

Keuning, D. (1998) *Management: a Contemporary Approach* (London: Pitman).

Kitchener, M. (1997) 'Professional Role Change under Quasi-market Transformation: the case of UK hospital doctors', paper presented to British Academy of Management Annual Conference, Queen Elizabeth II Conference Centre, London, September.

Llewellyn, S. (1997) 'The Virtues of Accounting Visibilities? Using budgets as two-way windows in UK hospitals', paper presented to British Accounting Association Scottish Conference, University of Edinburgh, September.

Marnoch, G. (1996) *Doctors and Management in the National Health Service* (Milton Keynes: Open University Press).

Marnoch, G. and K. Ross (1998) 'Flags and Badges – legitimacy in the new NHS', working paper, Department of Management Studies, University of Aberdeen.

Marnoch, G., L. McKee and N. Dinnie (1996) *Clinical Directors: roles, responsibilities and change*, Final Report, Shinton, Economic and Social Research Council.

Marnoch, G., L. McKee and N. Dinnie (1997) 'The Conditions of Power: The Case of Clinical Directors as Devolved Management in the NHS', conference paper presented to *Modes of organisation: power/knowledge shifts*, University of Warwick, April.

McKee, L. (1988) 'Conflicts and Context in Managing the Closure of a Large Psychiatric Hospital', *Bulletin of the Royal College of Psychiatrists*, 12(8), 310–19.

North, N. (1995) 'Alford Revisited: the professional monopolisers, corporate rationalisers, community and markets', *Policy and Politics*, 23(2) 115–25.

Ong, B.N., M. Boaden and S. Cropper (1997) 'Analysing the Medicine–Management Interface in Acute Trusts', *Journal of Management in Medicine*, 11(2), 88–95.

Pettigrew, A., E. Ferlie and L. McKee (1992) *Shaping Strategic Change: making change in large organizations, the case of the National Health Service* (London: Sage).

Scottish Office (1997) *Designed to Care* (Edinburgh: HMSO).

Strong, P. and J. Robinson (1988) *New Model Management: Griffiths and the NHS*, mimeo, Nursing Policy Studies Centre, University of Warwick.

Strong, P. and J. Robinson (1990) *The NHS: under new management* (Milton Keynes: Open University Press).

Thorne, M.L. (1997) 'Myth Management in the NHS', *Journal of Management in Medicine*, 11(3), 168–81.

Willcocks, S. (1997) 'Managerial Effectiveness in the NHS: a possible framework for considering the effectiveness of the clinical director', *Journal of Management in Medicine*, 11(3), 181–90.

9 Variations on a Theme: clinicians in management in England and the Netherlands

Bie Nio Ong and Rita Schepers

INTRODUCTION

The impact of health care reforms on the organisational behaviour of hospital doctors is a theme which is attracting considerable attention. Within many of the comparative health care analyses, the changing role of doctors tends to be discussed as an area of key importance (for example, Ham, 1997; Raffel, 1997) and research is being carried out that has as its central focus the medical profession's adaptation to policy changes in different health care systems (van Herk, 1997). Two main influences can be discerned which shape the current debates. First, the literature on the medical profession (Freidson, 1970; 1984; Döhler, 1989; Hafferty and Light, 1995; Larkin, 1995). One of the issues highlighted concerns the form and nature of self-regulation and the importance of clinical autonomy as a mechanism to safeguard professional control. Second, the shift in many countries towards a more managerially oriented health care system has drawn attention to the question of what this signifies for the medical profession: on the one hand, the economic power of managers over doctors is being considered the overriding theme (for example, Flynn, 1992) and, on the other hand, the argument is advanced that, at the micro level, doctors are difficult to control and thus continue to determine the shape of health provisions (Hunter, 1994).

The role of the medical profession within health care organisations across Western countries has not, as yet, reached a stable position, but it is possible to discern a number of trends which indicate the future direction of the relationship between doctors and managers. In this chapter we will focus, in particular, on the developments within the acute hospital sector in England and the Netherlands. The main reason for selecting these two countries is that doctors are explicitly drawn into managerial decision making because of organisational changes in the structure of health care, and because their professional expertise is used, through the mechanism of 'evidence-based care', in financial decision making. At the same time, important differences need to be considered in terms of the status of doctors and their relationship with third-party payers.

We focus on the acute sector. The pressure to deliver care of an explicitly stated amount and quality forces hospitals, and the doctors working within them, to develop systems which can enumerate activity and measure care processes and outcomes, and this implies greater accountability for medical decision making. The doctor can no longer operate as an individual agent in modern hospitals. Thus the relationships both between the individual doctor and the organisation, and between the doctors as a professional body and the organisation have to be scrutinised. This chapter draws on empirical research carried out in English and Dutch hospitals, examining the changing position of doctors and their representative bodies within hospital organisations. While comparisons are drawn between the two countries, the emphasis is placed on the Dutch situation, as the research data are more detailed and recent.

Briefly the English data were collected in a 650-bedded general hospital, organised in nine directorates, during 1994–6. The research design was an in-depth case study focusing on two directorates and using the other seven as comparative backdrop. Two other trusts were examined in less depth to provide a broader context. This study has been reported elsewhere (Boaden *et al.*, 1995; Ong *et al.*, 1997). The Dutch research was carried out in 1997 and covered nine hospitals across the country. In-depth interviews were carried out with medical managers and senior executive managers ($N=24$) supported by interviews with four policy analysts. All the material was analysed using the QSR NUD*IST computer-based tool and some of the findings have been reported by Ong and Schepers (1998).

First, the organisational structure of English and Dutch hospitals will be discussed with particular reference to the place of doctors within this structure (see Figures 9.1 and 9.2). This discussion will address the managerial characteristics of the medical role. Second, the manner in which doctors organise themselves as professionals within organisations indicates the routes and mechanisms that allow doctors to protect the medical domain, that is their expertise and esoteric knowledge as their power base. Third, we analyse how relationships within hospitals develop, most notably between medicine and management. Finally, we hope to signpost some implications for organisational behaviour.

HOSPITALS IN THE 1990s

The history of involving doctors in management in England has been chequered (Scrivens, 1988; Marnoch, 1996) with a major impetus occurring as a result of the 1990 reforms which introduced the separation of purchasers and providers. Predating the reforms were a number of experiments which drew doctors into managerial decision making within hospitals through the mechanism of clinical directorates (Chantler, 1990). The influence of these experiments, coupled with the broader policy shifts, stimulated many hospitals to review their organisations and, particularly, to examine the role of doctors within them. Because the contracting process moved central stage, doctors' knowledge about the form and content of clinical activities became an essential element in determining negotiations between hospitals and purchasers (health authorities and GP fundholders). This knowledge had to be harnessed in a structural manner so that the operational and strategic input of clinicians could be secured.

In the English model the clinical directorate constituted the cornerstone of the new hospital structure (Fitzgerald, 1994; Corbridge, 1995). A national survey on clinical directors (Simpson and Scott, 1997) described the content of the clinical director's job as perceived by 702 doctors. This survey confirmed findings from other studies in terms of the tension between the clinical and managerial work (for example, Dopson, 1994; Boaden *et al.*, 1996) but, notably, it emphasised the disenchantment felt by many clinical directors in terms of the amount of time they devoted to management tasks and the inverse value that was placed upon that work by colleagues and managers alike. The authors concluded that the involvement of doctors in management remained vulnerable and tenuous, with doctors indicating that they would either return to full-time clinical work (44 per cent) or retire from the health service (23 per cent).

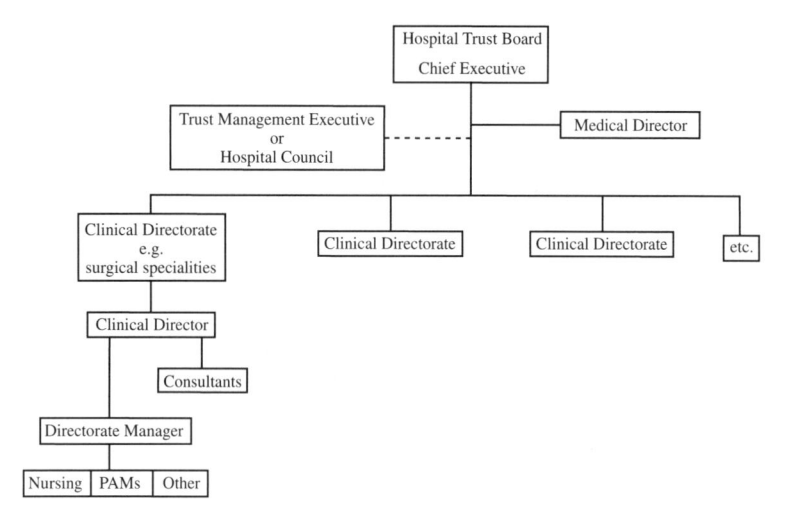

```
                    ┌─────────────────────┐
                    │ Hospital Trust Board│
                    │   Chief Executive   │
                    └─────────────────────┘
         ┌──────────────────┐         │         ┌─────────────────┐
         │ Trust Management │         │         │ Medical Director│
         │    Executive or  │─ ─ ─ ─ ─┤         └─────────────────┘
         │  Hospital Council│         │
         └──────────────────┘         │
┌──────────────────┐   ┌──────────────────┐   ┌──────────────────┐
│Clinical Directorate│  │Clinical Directorate│  │Clinical Directorate│ etc.
│      e.g.          │  └──────────────────┘  └──────────────────┘
│surgical specialities│
└────────────────────┘
     │
┌──────────────────┐
│ Clinical Director│
└──────────────────┘
     │      │
     │      └─── Consultants
     │
┌──────────────────┐
│Directorate Manager│
└──────────────────┘
  │
 Nursing | PAMs | Other
```

Notes:
1 TME/Hospital Council tends to include all clinical directors, other senior consultants and senior managers.
2 PAMs: professions allied to medicine.
3 There are, of course, variants of this model.

Figure 9.1   English hospital structure

The creation of boards overseeing hospital trusts included a medical director as a member of the executive team within the board. This role had primarily a strategic flavour through securing a medical input at the highest level of the organisation, and commonly responsibilities of medical directors included workforce planning, research and development (including evidence-based care) with medical audit, major clinical developments (such as cancer services) and risk management. In only a minority of cases did the medical director cease to carry out clinical work because the clinical involvement continued to be seen as providing the medical director with necessary credibility as a professional. Yet most medical directors understood that their position was primarily defined by its management content, and consequently by its allegiance to the board. Ashburner and Fitzgerald (1996) emphasised the 'boundary role' of medical directors where the latter acted as a bridge between the domains of management and medicine. As with clinical directors, this role created considerable tension, which could be further heightened by a sense of isolation because of a growing distance from medical colleagues.

The explicit management role of doctors in English hospitals is thus enshrined within management structures which define the locus of control for doctors at the level of clinical director (that is, the local, departmental/disciplinary level) and at board level in the post of medical director. While this formalised approach clearly defines the parameters of medical management, the reality is dependent on the motivation of individual doctors to be involved. The pressures upon medical managers cannot be underestimated because they inhabit a world in which there is, as yet, no new 'currency' defining criteria of success which are accepted equally by medical and managerial colleagues.

In the Netherlands hospitals began to change in 1983, when hospital budgeting was introduced. With hospital doctors (commonly called specialists) operating as independent contractors who predominantly were paid on a fee-for-service basis, the tension between the interests of the organisation and those of individual entrepreneurs was considerable. Furthermore hospital doctors were organised in a so-called 'maatschap' structure (a partnership of hospital consultants in which formal financial, service and quality arrangements have been agreed between the partners) which broadly meant that they worked in unidisciplinary partnerships that were based upon formal relationships (financial, workload and quality assurance) between the partners. Thus, from the mid-1980s, hospitals were developing more as corporate entities delivering care to explicit budgets agreed with insurers and government, while the specialists continued to work on an individualistic basis and were not fully integrated within the hospital structure (apart from the small minority of salaried doctors). At the same time, the government exerted considerable pressure on specialists by macro capping: that is, setting limits to the fees specialists could charge for their services (Scholten *et al.*, 1998). This policy of reducing the fees charged created a paradoxical situation: hospitals were working to an agreed budget. The specialists based within them wanted to maintain their income level and when fees were reduced they increased activity in order to achieve their previous income. This rise in activity caused the hospital budgets to run out of control, having a negative effect on total health care spending.

In response to spiralling health care costs, the Dutch health care reforms of the late 1980s introduced a competitive framework which is

122   *Clinicians in England and the Netherlands*

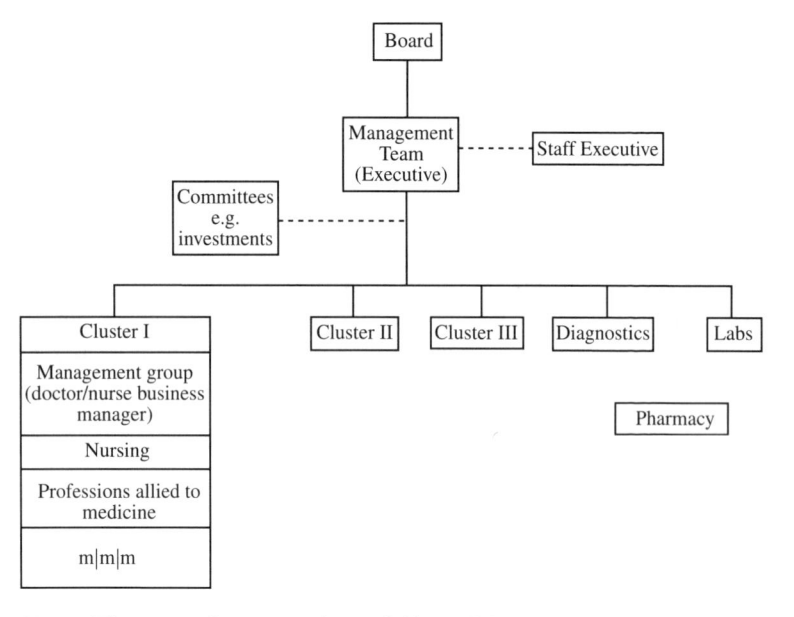

```
                              ┌───────┐
                              │ Board │
                              └───┬───┘
                                  │
                          ┌───────────────┐
                          │  Management   │        ┌─────────────────┐
                          │     Team      │--------│ Staff Executive │
                          │  (Executive)  │        └─────────────────┘
                          └───────┬───────┘
         ┌────────────┐           │
         │ Committees │           │
         │    e.g.    │-----------┤
         │investments │           │
         └────────────┘           │
   ┌─────────────────┬─────────┬──┴──────┬─────────────┬──────┐
┌──┴──────────────┐ ┌┴────────┐ ┌┴────────┐ ┌──────────┐ ┌────┴┐
│    Cluster I    │ │Cluster II│ │Cluster III│ │Diagnostics│ │Labs │
├─────────────────┤ └─────────┘ └─────────┘ └──────────┘ └─────┘
│ Management group│
│(doctor/nurse bus.│                                ┌──────────┐
│     manager)    │                                │ Pharmacy │
├─────────────────┤                                └──────────┘
│     Nursing     │
├─────────────────┤
│Professions allied│
│   to medicine   │
├─────────────────┤
│     m|m|m       │
└─────────────────┘
```

Note: There are, of course, variants of this model.

Figure 9.2   Dutch hospital structure

now beginning to develop (WVC, 1988). The recurring fee reductions, coupled with the new policy framework, necessitated a rethink of the relationship between hospitals and doctors. Thus the payment system could not be seen as separate from the management of hospitals. In 1989 agreement was reached at the highest level to implement a macro budget for specialist medical care at the 1989 level, which would be in place until 1993 (Scholten *et al.*, 1998). This created a degree of stability, which could then reinforce the structural proposals put forward by the Commission, 'Modernisation of curative care' (Commissie, 1994), which argued that the modern hospital should operate as an integrated medical-specialist organisation (the so-called 'Biesheuvel' model). Initially this meant that doctors were subordinate to management, but the model has evolved towards collaboration between specialists and hospital. This tendency will be further strengthened by a new Act formally establishing the integration of hospitals and specialists (to be introduced in 1999).

At present there is a shift in focus towards local structures, opening the way for so-called 'local experiments' (introduced by the secretary of state in 1994) to be implemented, whereby hospitals, insurers and medical specialists come together in collaborative working arrangements around patient-centred care.

It should be noted that the local experiments are not uniformly accepted and successful across the country. This is also due to the various interpretations of the medicine–management relationship whereby the Biesheuvel model is arguably not necessarily the best approach for getting doctors interested in management (Scholten, 1996). However the local experiments in themselves provide opportunities to restructure hospitals and the role of specialists within them. This is borne out by a large number of hospitals now 'overturning' their organisation by re-emphasising the primary process of care and using that as its organising principle. Specialists tend to head units, and are supported by nursing and general managers; many hospitals group units into clusters or sectors in order to create more cohesion between them and to facilitate working across the hospital as a whole. Again clusters are headed by specialists, so that doctors are directly involved with operational management (at unit level) and strategic management (at cluster level). At the board level each hospital has a medical director (who is a full-time manager) alongside the general manager (the majority of Dutch hospitals have a two-member executive team) symbolising the difference between the subordination of doctors to management (the Biesheuvel model) and the collaborative relationship, based upon equality between doctors and managers.

The tensions experienced by English medical managers are very similar to those of their Dutch counterparts. They are also searching for an appropriate balance between medicine and management, at both the philosophical and the implementation level. In both countries doctors 'volunteer' to become managers, but maintain a degree of clinical involvement in order to continue to be credible to clinical colleagues, and to have a fall-back position. The demands placed upon those doctors by operating in two domains have been well documented (for example, Dopson, 1995). The issue, however, is to conceptualise a different role for medical managers, not one which needs to straddle, but one which integrates, medicine and management. Before we address this question, we would like to turn to the way in which doctors organise themselves within hospitals, as this largely determines their perception of themselves as a professional group – and as individuals within it – and their relationship with management and third parties (health authorities or insurers).

MEDICAL ORGANISATIONS IN HOSPITALS

The clinical directorate structure in English hospitals represents a managerial model which aims to draw doctors into the operational, and to some degree the strategic, management of the organisation. At the same time, consultants operate within professional groupings which, although not directly integrated into management structures, exert an important influence on decision making at all levels. As clinicians from a particular discipline they are invariably members of the relevant Royal College which defines standards for skills and expertise, as well as quality. Royal College guidelines are strongly promoted from the outside, but equally from inside the organisation, and place considerable pressure on management decisions regarding, for example, maintaining training status through the right skill mix and number of consultants. Medical power can be maintained through this route because the judgement of medical knowledge and practice stays firmly within the hands of the profession and it is a brave manager who will go against advice which has Royal College backing. At the same time, criticism is growing (from managers and doctors alike) of this type of standard setting as an absolute benchmark, separate from the reality of financial constraints and rising workloads. For example, the combined demands imposed by the reduction of junior doctors' hours and continuing medical education for senior clinicians mean that many clinical directorates have considerable difficulties achieving, for example, Royal College standards of supervision and formal teaching of junior staff.

Within hospitals consultants can be organised in various ways, but most commonly they come together as medical professionals (not divided by disciplinary boundaries) in groupings such as the Medical Advisory Committee or the Trust Management Executive. Ultimate responsibility (and thus decision making) will continue to rest with the executive team and the board, although in most hospitals the intention is to draw clinicians into the organisation through formulating corporate objectives, assessing current strengths and weaknesses and setting out strategic plans. For this to work in practice a sophisticated understanding of clinical management has to be developed whereby boundaries between clinical directorates are blurred and the 'corporate good' takes precedence. This, of course, depends to a large extent on the quality of the evidence available about cost-effectiveness and outcomes of treatments (Sackett et al., 1996) so that appropriate comparisons can be made across the directorates. In the absence of robust evidence, agreement on intraorganisational priorities are hard to establish and the 'shroud waving' of powerful clinicians is capable of continuing to determine organisational objectives.

The Dutch hospital doctors have a chequered history with regard to their organisational capacity as professionals and it is beyond the scope of this chapter to detail the splits and disputes that have taken place over the last two decades. Suffice it to say that hospital specialists tend to be organised in a National Specialist Association which has its own subgroups relating to disciplines, while the medical umbrella organisation brings GPs and specialists together. These bodies have important roles in relation to national negotiations on policy and pay and conditions.

The focus of this chapter is the examination of the way doctors are organised within hospitals. In the Dutch context a number of different forms exist alongside each other. At the lowest level is the 'maatschap' which represents possibly the strongest form of organisation and allegiance for specialists. Their individual interests as medical professionals are intimately connected with the interests of the maatschap, and these are not always in harmony with either the unit/cluster or the hospital interests. The question of identity plays an important role and this is clearly expressed by a specialist manager of a large hospital:

> In the maatschap you have to show that you are straight. If you want to be the leader of a maatschap or a department you have to show that you are not a vassal of the management. That was one of the worries in my case at the beginning. If you want to come out of the maatschap and become a Speciality Manager, you become a lackey of the hospital.

Allegiance to the maatschap is considered paramount and other colleagues have been closely watching this speciality manager as to whether he let hospital interests prevail. For him the key issue is to align the two sets of interests and to convince his partners that, when the hospital does well, this has a positive impact on the maatschap.

The second professional organisation level is through the staff convent which is based upon the concept of collegiality. Each hospital has its own convent which consists of all specialists, and representatives of all the 'maatschappen' (partnerships) form the staff executive. Thus the size varies according to the number of maatschappen, and the members take part on a voluntary basis. The strength of this executive is highly dependent on the calibre of the people, the time they can devote to the work and the support from their own maatschap. Furthermore the issue of representation is not wholly clear-cut as there is no agreed model for proportional representation (that is, a bigger maatschap having more members on the executive) or a fixed number of representatives per maatschap (Moen, 1993). It is important to note that specialists have contracts with hospitals which therefore create an exclusive and interdependent relationship.

Currently the key debate centres around the role of the staff convent and staff executive within the new hospital structure. Research on management participation in Dutch hospitals (Versluis and Hesselink, 1994) outlined a number of issues, including the relationship between staff and staff executive, the question of giving a mandate to the staff executive, and the staff executive as advisor to the executive team or as participant in management decision making.

In the present situation hospital managers, and in particular the executive team, have regular contact with the staff convent and discuss both strategic and operational issues. The survey carried out by Versluis and Hesselink (1995) demonstrates that 75 per cent of specialists feel that the staff executive has to discuss all organisational decisions with the staff convent. In reality, this happens only in about 50 per cent of cases, and the survey findings illustrate that specialists are of the opinion that more time and effort has to be devoted to communication with the 'rank-and-file'. A staff chairman reflected on this question as follows: 'Most staff members feel that management should not cost them too much energy. They delegate that to the staff executive. But they always want to know precisely what the executive does, and we have to brief them extensively, preferably in writing.'

Emphasis is placed on communication both between staff and their executive and between individual staff members in order to ensure that there is openness and 'ownership' of decisions that are taken into the management arena, yet there are real problems in getting everyone on board, as explained by a medical manager: 'The other big problem is that we are so large that you have staff meetings with only half the people present. The silent majority you never know. Then the die-hards suddenly come in and they thump on the table again.'

The problem of involvement across the whole organisation is a familiar one, but it presents a particular dilemma if the individuals and their partnership have a high degree of autonomy. Making decisions within a staff convent that has assembled a minority offers only a fragile foundation, and the staff executive will have to seek other avenues for building more robust communication and decision making based upon consensus.

The second question follows on from the above: namely, whether and to what extent the staff executive has a mandate. The study by Versluis and Hesselink (1995) explicitly addresses this question and they argue that staff executives are rarely given a full mandate by their rank-and-file, for a number of reasons: first, the body of specialists do not form a homogeneous group; second, the interests of the various maatschappen vary, in particular with regard to protection of income; third, doctors continue to be guided by the idea of professional autonomy and they resist being bound by decisions which can have a direct impact upon individual practice, as for example through the introduction of clinical guidelines (van Herk, 1997).

It can be argued that both in the UK and in the Netherlands protectionism is under threat because of the shift to cost-effectiveness, which forces medical decision making and practice to be more transparent and accountable. This gives rise to a number of tensions which are clearly illustrated in the Dutch case. One example is drawn from an interview with a staff chairman of a medium-size hospital who had to contend with the problems discussed in the Versluis and Hesselink study. The chairman analysed the situation as follows:

> The dynamics are that giving the mandate to the staff executive is all lopsided. The fact that when you have a mandate you have power is seen as problematic by the staff. Secondly, when you get involved in management you move towards the board, and the staff finds that very threatening. Thirdly, the staff chairman and secretary are freed up for two days a week. You become knowledgeable, you follow courses, you read, you get a broader overview and you start thinking differently. They feel that. You acquire power through knowledge because you do things they don't. You can try to eliminate that through communication, but the feeling persists.

The staff body does not want to give a full mandate because they like to keep a degree of control over their staff executive, which they see as becoming 'experts' in management. This growing insight into managerial decision making is often interpreted as moving over to 'the other side' and it is very important that a visible loyalty to the staff's perspective is demonstrated. In this case the staff chairman exacerbated his precarious position by insisting on the structural model of the staff executive being part of management, thus creating a distance between the specialist body and the staff executive that was perceived by the rank-and-file to be greater than the distance between the staff executive and the executive team.

In practice, the problem is twofold: first, how the staff executive as representing specialists should define itself as either a professional interest body or as a participant in management; and second, how this affects the mandate given to the staff executive whereby control is strongly maintained by the rank-and-file (aligned to professional interests) or where they enjoy considerable freedom to be responsive and involved in strategic decision making (moving towards management). There is discussion in the Netherlands about the appropriate models and the comparison between doctors being involved in operational or strategic management (see, for example, Schepers *et al.*, 1996).

The discussion thus far has outlined the various organisational forms that doctors adopt within English and Dutch hospitals, but has not been able to state unequivocally which approach allows medical expertise and knowledge to be protected while not becoming so protectionist as to obstruct management and decision making of the hospital as an organisation. In order to begin to create new organisational forms we have to look at the developments in the relationship between management and medicine at hospital level.

DOCTORS AND MANAGERS: PARTNERS OR PROTAGONISTS?

With regard to the changes in English hospitals, we have argued previously (Ong *et al.*, 1997) that an adaptive model is emerging as a result of changing health policy. The oppositional relationship between doctors and managers is weakening and dynamic alliances between the two parties are developing because both are realising their increased interdependence under the influence of policy pressures such as contracting by clinical pathways or achieving cancer accreditation.

The comparisons with the Dutch situation, where policy initiatives have fundamentally reshaped relationships, are interesting. The most recent and important change is the introduction of the local experiments which bring insurers, specialists and hospitals together in negotiations about service configurations, resources and longer-term plans. The impact of this policy is immediate and one staff chairman voiced this clearly: 'We have now the tripartite negotiations and that has partly contributed that the interests of the doctors and the hospital are in the same direction. In the past the doctors wanted to grow and the hospital wanted to stop. Now it is parallel.' This change of opinion could be explained as due to structural changes and the different way in which negotiations are framed, but also because the immediate financial interests of both the hospital and the specialist were at the centre of the negotiations. The success of the hospital and the individual specialists and their 'maatschappen' were shown to be mutually reinforcing, thus underlining the concept of interdependence.

From the perspective of the executive team the advantages of this convergence are obvious. The chief executive of a large hospital explained this, with reference to bridging management and medicine:

This is the most important shift: that you are not thinking in two separate worlds. There has to be reciprocal influence. At the moment that this happens between doctors and management and other doctors at the level of units, you suddenly see that things are moving along better.

In this case, the local experiment was not the prime force behind management and medicine moving closer together, but the specific historical event of creating this new hospital from a merger of three. Not only did it become possible to build a brand-new building with considerable input from the medical staff, but it also allowed a 'clean-up' of the middle management cadre and the freeing of resources to support and strengthen clinical management at unit and sector level. The redefinition of the whole organisation, almost like beginning with a clean slate, allowed for innovation in the management–medicine interface.

It would be easy to dismiss this case as unrepresentative, because of the fortuitous local circumstances, but it highlighted important issues.

First was the considerable investment in selection and training of doctors for management. Doctors had to apply for the post of specialist manager and, on appointment, received structured training as a group, and in some cases personal coaching. Secondly a support structure was created with management back-up for the staff executive in the form of a medical co-ordinator (facilitation, organisational development, liaison between 'maatschappen'/units and management). Within the clusters general managers were working alongside cluster chairmen (who were doctors), and within the units general management support was put in place in the form of a unit manager and *ad hoc* assistance from central departments such as finance or personnel. Thirdly decentralisation was far advanced, with budgets being devolved, and thus accountability for business plans and their implementation became more transparent. Responsibilities and freedoms were increasingly clearly defined, but the balance with corporate decision making had to be maintained. This was most clearly articulated through the 'investment board', which is chaired by the chief executive, with representation from the clusters and all medical disciplines. Here decisions are made about new developments based upon clinical and cost-effectiveness evidence and advice from external experts. The corporate perspective has to prevail over cluster or unit interests and the test of whether management and medical thinking are coming closer together is most prominent within this board.

Looking at the discussions nationally, the concept of 'being equal partners' is considered of crucial importance. This is translated into finding a way at the strategic and operational level of allowing the medical and management perspectives equal weight. Most people interviewed felt that this concept served as a guiding principle for the way working relationships should be developed. There were also some sceptical voices, for example, a specialist manager who had been prominent in the national medical body said: 'The majority of doctors is hardly equipped to operate as equal partners, to function in that way at board level. They work medically orientated because they have no other expertise.' Yet this perspective should not be seen as dissonant, because it underlines the importance of the change in culture which is required on the part of both doctors and managers. Moen and de Roo (1997) argue that many managers and doctors are moving towards a new negotiated order which clarifies their respective roles and responsibilities. While this can lead to segmentation, it also affords the opportunity to define hybrid and more flexible roles which straddle the two paradigms.

CONCLUSION

Merging the medical and management paradigms is an unfinished process in both countries. In England there is considerable uncertainty about the sustainability of the medical manager model with the doctor as a practising clinician (Simpson and Scott, 1997). The 1997 white paper, *The New NHS – Modern and Dependable* outlines the concept of clinical governance. With explicit accountability for quality of care at the level of the chief executive, the prominence of clinical input into the management process becomes more pronounced.

The Dutch case demonstrates that a number of factors influence changes within hospitals. The external environment includes policy makers who define the parameters such as financial allocations, the local experiments and the degree and level of competition. At a more local level, hospitals and specialists relate to the external world in the shape of insurers, and increasingly operate with a common voice vis-à-vis the insurers. The creation of investment boards demands that specialists supersede territorial concerns and begin to think more corporately. Comparing the developments in the Netherlands and in England, the broader policy pressures, coupled with the internal shifts in hospital organisations, continue to push for collaborative working between doctors and managers. For the foreseeable future the formulation and testing of organisational models which articulate the input of clinicians in strategic management will remain an important issue. The opportunities for bringing the clinical quality and cost-effectiveness agendas closer together need to be grasped if medicine and management are to work towards shared goals.

References

Ashburner, L. and L. Fitzgerald (1996) 'Beleaguered professionals: clinicians and institutional change in the NHS', in H. Scarbrough (ed.), *The Management of Expertise* (London: Macmillan).
Boaden, M., B.N. Ong and S. Cropper (1995) *Clinicians' involvement in the management of Trusts* (Cheadle: British Association of Medical Managers).
Chantler, C. (1990) 'Management reform in a London hospital', in N. Carle (ed.), *Managing for Health Results* (London: King Edward's Hospital Fund).
Commissie Modernisering Curatieve Zorg (1994) *Gedeelde zorg, betere zorg*, Rijswijk.
Corbridge, C. (1995) 'Pandora's box: clinical directorates and the NHS', *Journal of Management in Medicine*, 9(6), 16–20.

Department of Health (1997) *The New NHS – Modern and Dependable* (London: HMSO).

Döhler, M. (1989) 'Physicians' professional autonomy in the welfare state: endangered or preserved?', in G. Freddi and J. Björkman (eds), *Controlling Medical Professionals* (London: Sage).

Dopson, S. (1994) 'Management: the one disease consultants did not think existed', *Journal of Management in Medicine*, 8(5), 25–36.

Fitzgerald, L. (1994) 'Clinical management: the impact of a changing context on a changing profession', in J. Leopold, M. Hughes and I. Glover (eds), *Beyond Reason? The National Health Service and the limits of management* (Aldershot: Avebury).

Flynn, R. (1992) *Structures of Control in Health Management* (London: Routledge).

Freidson, E. (1970) *Profession of Medicine: a study of the sociology of applied knowledge* (New York: Dodd Mead).

Freidson, E. (1984) 'The changing nature of professional control', *Annual Review of Sociology*, 10, 1–20.

Hafferty, F. and D. Light (1995) 'Professional dynamics and the changing nature of medical work', *Journal of Health and Social Behaviour* (extra issue), 132–53.

Ham, C. (1997) *Health Care Reform: Learning from international experience* (Buckingham: Open University Press).

van Herk, R. (1997) *Artsen onder druk* (Utrecht: Elsevier/de Tijdstroom).

Hunter, D. (1994) 'From tribalism to corporatism: the managerial challenge to medical dominance', in J. Gabe, D. Kelleher and G. Williams (eds), *Challenging medicine* (London: Routledge).

Larkin, G. (1995) 'State control and the health professions in the UK: Historical perspectives', in T. Johnson, G. Larkin and M. Saks (eds), *Health Professions and the State in Europe* (London: Routledge).

Marnoch, G. (1996) *Doctors and Management in the National Health Service* (Buckingham: Open University Press).

Moen, J. (1993) 'Samenwerking tussen maatschappen, stafbestuur en management: Beleids- of managementparticipatie?', *Ziekenhuis Management Magazine*, 19, 198–201.

Moen, J. and de Roo A. (1997) 'Het samenspel tussen specialist en ziekenhuis: co-makership', *Medisch Contact*, 52, 27/28, 871–73.

Ong, B.N., M. Boaden and S. Cropper (1997) 'Analysing the medicine–management interface in acute trusts', *Journal of Management in Medicine*, 11(2), 90–7.

Ong, B.N. and R. Schepers (1998) 'Comparative perspectives on doctors in management in the UK and the Netherlands', *Journal of Management in Medicine*, 12(6), 378–390.

Raffel, M. (ed.) (1997) *Health care reforms in industrialised countries* (Pennsylvania: Pennsylvania State University Press).

Sackett, D., W. Rosenberg, J. Gray *et al.* (1996) 'Evidence-based medicine: what it is and what it isn't', *British Medical Journal*, 312, 71–2.

Schepers, R., N. Klazinga and G. Scholten (1996) 'Beter maten dan managers. Managementparticipatie in de Nederlandse ziekenhuizen', *Gezondheid*, 4(1), 30–8.
Scholten, G. (1996) 'Boekbespreking van G. Visser e.a. Ondernemend besturen: ziekehuismanagement van overmorgen', *Gezondheid*, 4(3), 355–8.
Scholten, G., A. Roux and J. Sindram (1998) 'Cost control and medical specialist payment. The Dutch alternative', *International Journal of Health Planning and Management*, 13(1), 69–82.
Scrivens, E. (1988) 'The management of clinicians in the NHS', *Social Policy and Administration*, 22, 24–34.
Simpson, J. and T. Scott (1997) 'Beyond the call of duty', *Health Service Journal*, 8 May, 22–4.
Versluis, J. and M. Hesselink (1995) *Managementparticipatie van medisch specialisten en decentraal organiseren* (Utrecht: LSV, NVZ, NZi).
WVC (1988) *Willingness to change* (Rijswijk: Ministerie voor Welzijn, Volksgezondheid en Cultuur).

10 Leadership in the NHS: what are the competencies and qualities needed and how can they be developed?

Beverly Alimo-Metcalfe

INTRODUCTION AND BACKGROUND

Several commentators have documented the organisational reforms that the NHS has passed through since its inception in 1948. Each has supposedly been introduced to enable the NHS to meet the increasing demands to become more efficient and effective in delivering health care (for example, Harrison *et al.*, 1992; Paton, 1996; Stewart, 1996). In line with these demands, academics and practitioners in the field of management development have responded enthusiastically and suggested various interventions aimed at supporting the service with techniques and processes for developing managers' skills and other capabilities to enable them to cope with the ever-increasing complexity of the environment in which they operate.

Such academics and practitioners have been informed by the vast literature that has been produced in the area of 'leadership', and of organisational behaviour, in particular. Most conspicuous in its influence is the development in the 1980s, led mainly by US researchers, of the two complementary models of transactional and transformational leadership (for example, Bass, 1985, 1990; Bass and Avolio, 1984). At a very simplistic level, the distinction between these models marks the difference between management and leadership. The management, or transactional leadership, model places emphasis on organising and planning the use of resources, 'fixing' problems that emerge and monitoring the progress of activities directed at achieving predictable outcomes and predetermined objectives. On the other hand, 'leadership', or transformational leadership as it is now called, goes beyond management and involves creating new scenarios and visions, challenging the status quo, initiating new approaches and exciting the creative and emotional drive in individuals to strive beyond the ordinary to deliver the exceptional (Bass *et al.*, 1996, pp.6–7). Figure 10.1 shows the dimensions of transformational leadership identified by Bass (1985).

The case for more leadership in the NHS is the same as that exalted in most current organisational literature, namely the need to deal with greater uncertainty and increased complexity, the constant need to be able to adapt to change, in the context of limited resources being available (for example, Boot *et al.*, 1994; Morgan, 1993). The validity of the transformational leadership model has been supported by growing evidence from studies which have investigated its efficacy. Bass *et al.* (1996) have produced summaries of the relevant literature and concluded that research has shown that transformational leadership correlates strongly with a variety of objective outcome measures, including the following:

- commitment, effort, performance and job satisfaction of followers;
- employee innovation, harmony and good citizenship;
- financial performance of organisations;
- performance in the public sector.

Charismatic:	highly esteemed, role models whom followers strive to emulate who align others around a vision, common purpose and mission
Inspirational:	provides meaning and optimism about the mission and its attainability
Intellectually Stimulating:	encourages followers to question basic assumptions, and to consider problems from new and unique perspectives
Individually Considerate:	works with followers, diagnosing their needs; transcends their self-interests, enhances their expectations and develops their potential

Figure 10.1 Transformational leadership: the new leadership paradigm
Source: B.M. Bass, B.J. Avolio and L. Atwater (1996) 'The transformational and transactional leadership of men and women', *Applied Psychology: An International Review*, 45(1) 5–34.

Transformational leadership has been found to produce comprehensively more impressive results than transactional leadership. However it should be added that transformational leadership does not replace a transactional leadership approach but augments it, since both are required in complex organisations (Bass *et al.*, 1996). Whilst this new paradigm of transformational leadership appears promising in informing management development and selection processes in organisations such as the British National Health Service, before assuming its validity in the UK public sector context there is a need to investigate what are perceived to be the components of transformational leadership in the NHS, and indeed what we have learned about the most important management competencies perceived to relate to notions of effectiveness. This chapter will examine some relevant research emerging from studies conducted in the NHS and will make recommendations as to how the service might move to a model of good practice in selecting and developing for leadership and management in the future.

WHY THE CASE FOR MORE LEADERSHIP IS SO PRESSING FOR THE NHS

It has long been known that professionals in the NHS experience high levels of stress. For example, it has been found that doctors, pharmacists and other therapists are twice as likely to commit suicide as those in the general population (Charlton *et al.*, 1993); morbidity of health workers is also high, with particular risks of drug and alcohol abuse (Heim, 1991); there is increased frequency of stress-related illness in health professionals exposed to the upheavals of health service reforms (UNISON, 1994; Caplan, 1994); and applications for early retirement on grounds of psychological stress jumped from 17 per cent in 1994/5 to 23 per cent in 1995/6 (*Lancet*, 1994; Moore, 1996).

Fortunately the situation has been taken seriously by the Department of Health, which funded a longitudinal research project to look at the state of mental health of staff working in the NHS, including doctors, nurses, managers and ancillary staff (Borrill *et al.*, 1996). In its first year-end report, the researchers stated that 'the mental health of the workforce was found to be considerably poorer than that of employees working outside the NHS' and, somewhat alarmingly, that 'overall, 27% of staff working in the NHS Trusts are probable psychiatric cases' (p.1).

Interestingly the professional group with the highest proportion of highly stressed individuals appeared to be NHS managers.

Sustained high levels of stress can have a significant negative impact on various work behaviours, including satisfaction at work, turnover, self-esteem, and performance, and negative affects such as hostility, anxiety and depression (Golembiewski and Boss, 1992). As worrying as these findings are for the general population, they become a particularly strong cause for concern in the context of doctors. Firth-Cozens (1994) noted that overstressed doctors are more likely to make mistakes, to be less empathic to patients, to be less communicative with colleagues, to take patients' symptoms at face value and to become uninterested and disenchanted with their work. But what does this have to do with leadership?

In answering this question we should note that, in a recent study which reviewed the data emerging from employee climate surveys (which request anonymous feedback), the researchers concluded:

> organisational climate studies from the mid-50s to the present routinely show that 60%–75% of employees in any organisation, irrespective of industry, sector, level or occupational group, report that the worst and most stressful aspect of their job is their immediate boss. ... Good leaders may put pressure on their people, but abusive and incompetent management create billions of dollars of loss productivity each year. (Hogan et al., 1994, p.494)

The effect of leadership style on staff stress levels has received research attention, and several studies have found similar data to those emerging from a study by Offerman and Hellman (1996), namely that at least three key variables consistently emerge as contributing to dysfunctional stress in staff. One is concerned with the degree of performance pressure, which clearly will not decrease in organisations, at least in the foreseeable future. The other two variables are the degree to which an individual experiences the freedom and autonomy to control their work and use their discretion; and the extent to which work roles and objectives are clarified. The greater the discretion and clarity of objectives, the lower the experienced dysfunctional stress.

Studies which have compared the stress levels of groups which experience similarly high levels of performance pressure have found significant differences in perceptions of stress amongst the groups. Lower levels of perceived stress were found in groups in which individuals experience a high degree of autonomy and discretion (for example, Jackson *et al.*, 1986; Carrère *et al.*, 1991). One example of this is a study of various groups of nurses which were matched for the amount of job demand they experienced (Wall *et al.*, 1996). Those groups which experienced high job demand ('the amount of work required from the employee, the extent to which he or she has to work under time pressure, and the degree to which the employee is expected to complete conflicting job demands': Sargent and Terry, 1998, p.219) were then subdivided into groups experiencing various degrees of job control, from very high to very low. Their levels of job satisfaction were measured. The findings concluded that, even when controlling for high levels of job demand, the greatest job satisfaction was experienced by those who felt they were in situations which permitted high job control, and the least job satisfaction was experienced by those with very low levels of job control.

Other characteristics of the job which relate to psychological well-being at work include identity with the job, a sense that the job one is performing is valued, opportunity to use a variety of one's skills, and quality of feedback (for example, Kelloway and Barling, 1991). Notwithstanding the very consistent research findings which have found that these factors, together with job control, relate significantly to job satisfaction and perceptions of stress, managers seem relatively unaware of the role that they can play in influencing these variables which can either reduce or exacerbate stress. Apart from the very real costs to individuals of high stress in the job, such as physical and mental health (for example, Barling, 1990; Cooper and Marshall, 1976; Karasek *et al.*, 1981; Warr, 1990), there are considerable organisational costs, including those arising from lower performance and low job commitment. Despite the fact that stress is largely environmentally caused (for example, Golembiewski and Boss, 1992; Brookings *et al.*, 1985; Parker and DeCotiis, 1983; Sutherland and Cooper, 1988), organisations, including those such as the NHS which one might expect to be more concerned with employee well-being, appear relatively oblivious to the environmental factors which can create such a destructive outcome.

Offerman and Hellman (1996) state: 'the traditional approach to reducing stress places responsibility for stress management on the employee...organisations could realise more benefit by interventions designed to change the system producing the stress before the stress occurs...leader behavior is...amenable to change and thus is a prime target for efforts to alleviate unhealthy levels of work stress (p.389). They add: 'Managers with controlling styles, who fail to clarify organizational goals and responsibilities and who exert undue pressure have work groups with higher levels of stress' (ibid.).

It is clear that the leadership style which is most likely to alleviate such stress is one which is more empowering and which allows staff a greater degree of autonomy and discretion: the transformational approach. However this must be combined with the functional transactional behaviours of clear goal setting.

GENDER AND TRANSFORMATIONAL LEADERSHIP

Before looking at ways in which the development of transformational leadership can be supported, it is important to note that current research which has investigated the possible effects of gender on preferred leadership style has revealed consistent findings that women managers, in general, are more likely than men managers, in general, to adopt a predominantly transformational approach to leadership. Whether citing Rosener's study (1990), in which 300 or so senior female US managers and a matched group of 100 or so male senior managers were asked to describe their typical leadership style, or research based on interviewing senior females and males in the NHS (Alimo-Metcalfe, 1995) and local government (Sparrow and Rigg, 1993) to elicit their constructs of leadership, the findings emerge that women in general are more likely to cite the transformational style, and men the transactional.

It should be added that Roseners' research was criticised for the assumptions it made that the women would actually adopt the style they espoused. While such a criticism is legitimate, given the research, which reveals that, in general, managers' perceptions of their own leadership and management competence do not correlate significantly with the perceptions of their co-workers (a matter which will be discussed in a later section), Alimo-Metcalfe has argued that there are, nonetheless, serious implications of these findings when it comes to identifying criteria for the selection, promotion and development of managers (for example, Alimo-Metcalfe, 1994). But, more importantly perhaps, it should be added that there have also been studies which have investigated subordinates' descriptions of their managers' leadership style. Three such studies were conducted by Bass *et al.* (1996) who concluded that women managers are perceived to be significantly more transformational than male managers. A later section in this chapter will discuss NHS research with respect to gender and leadership.

HOW IS LEADERSHIP DEVELOPED?

Given the superiority of the transformational leadership model of managing over the transactional approach, organisations have reviewed their approaches to developing managerial excellence. One of the most important developments in this area has been the introduction of 360 degree or multi-rater feedback processes (MRF). With greater emphasis being placed on the need for managers to understand the impact of their behaviour on their staff's and colleagues' motivation and effectiveness, the traditional approach to providing managers with formal feedback through organisational appraisal systems is no longer regarded as sufficient, or even always appropriate.

Appraisal has been criticised on several grounds. Fletcher (1993) maintains that there is a general dissatisfaction with appraisal schemes used in organisations, partly because appraisal has become an emotive word (McBeath, 1990, cited in Fletcher, 1993), partly because it is associated with evaluation of performance which may be regarded as unfair and/or invalid; and because it is often conducted poorly. Further criticisms include the following:

- increasingly managerial effectiveness relates to success in managing teams, so feedback from team members would be more valuable (Fletcher, 1993);
- the typical appraiser is the manager's boss and she or he has limited opportunities to view the manager's behaviour (Redman and Snape, 1992);
- appraisal has traditionally been seen as a personnel-led process rather than managerially owned (Fletcher, 1993);
- the views of peers are seen as more relevant in providing task-relevant feedback, and more useful if developmentally focused rather than evaluative (Wohlers and London, 1989);
- managers are becoming increasingly aware of the value of feedback from a range of individuals both within and external to the organisation, including peers, subordinates, their supervisor/boss and 'clients' (Redman and Snape, 1992; Church, 1997).

The process of 360 degree/MRF involves managers rating themselves on certain competency-based or attitudinal dimensions, and asking work colleagues (typically their boss, plus some peers and some subordinates) to rate them anonymously. The outcome of the process is a report which usually shows the manager's self-ratings against the average ratings of their colleagues (see Alimo-Metcalfe, 1998, for a more detailed description of the process and research findings).

Why use 360 degree feedback?

The rationale for using 360 degree/MRF has included the following points:

- the data are possibly more valid since they include observations of subordinates who are in closer contact than the manager's superior to observe directly a wide range of management behaviours (Redman and Snape, 1992);
- this allows subordinates to observe how the manager reacts, not only in a crisis, but also in more normal day-to-day conditions (ibid.);
- 'subordinates' observations come from the receiving end of many managerial practices and this may give such feedback greater validity than under traditional top-down systems' (ibid., p.33);
- the use of multiple appraisers (for example, boss, peers, subordinates) includes perceptions from raters 'who have had opportunities to observe different aspects of a person's behaviours. The use of multiple raters increases the reliability, fairness and acceptance of the data by the person rated' (London *et al.*, 1990, p.18);
- it certainly signals that the organisation 'wants to encourage managers to emphasise the importance of effective management of people and positive relationships with peers within and between departments' (ibid., p.17);
- 'subordinates and peers' ratings are considered especially valuable by employees when ratings are used for developmental rather than evaluative purposes' (London and Wohlers, 1991, p.376);
- studies examining the effects of upward feedback have found that it leads 'to subordinates perceiving positive changes in the boss's subsequent behaviour' (Reilly *et al.*, 1996, p.608) and that improvement was sustained two years later (ibid.);

- subordinates were more satisfied with their managers when there was greater congruence between their perceptions of the manager and the manager's self-perception (Wexley *et al.*, 1980);
- 'Leaders suffering from inflation (of their effectiveness) overestimate their influence and are likely to misjudge and misdiagnose their own need for improvement' (Bass and Yammarino, 1991, pp.439–40);
- an inflated self-evaluation may result in 'career derailment' of managers as they become more arrogant and see little or no need to change (McCall and Lombardo, 1983).

Findings from 360 degree/multi-rater feedback (MRF) research

There now exist substantial data on the psychometric characteristics of 360 degree/MRF, which appear to be remarkably consistent. Briefly summarised, the findings include the following:

- self-ratings tend to be inflated (for example, Podsakoff and Organ, 1986; Harris and Schaubroeck, 1988; Atwater and Yammarino, 1992);
- subordinate ratings have been found to be more similar to supervisor (bosses) ratings than to self-ratings in terms of convergent validity and leniency effects (Mount, 1984, cited in Wohlers and London, 1989, p.376);
- self-ratings are less highly related to ratings by others (peers, supervisors or subordinates) than peers', supervisors', and subordinates' ratings are to one another (Harris and Schaubroeck, 1988, cited in Atwater and Yammarino, 1992);
- self-ratings are also less accurate than ratings from peers or supervisors when compared to objective criterion measures (for example, Hough *et al.*, 1983, cited in Atwater and Yammarino, 1992);
- inaccurate self-raters (that is, those with self-ratings that differ greatly from observer ratings) obtain lower performance ratings than their more accurate counterparts (for example, Bass and Yammarino, 1991; Flocco, 1969, cited in Atwater and Yammarino, 1992);
- the magnitudes of the correlations between predictors (including measures of ability and measures of experience) and ratings of leader behaviour, as well as between leader behaviour and performance

measures, vary as a function of the degree of self–other agreement (Atwater and Yammarino, 1992)
- self-ratings of leadership have been found to fail to correlate with performance measures and promotability, whereas the parallel subordinates' ratings were associated with these measures (Bass and Yammarino, 1991).

WHAT WOULD FORM THE BASIS FOR A LEADERSHIP DEVELOPMENT 360 DEGREE/MRF INSTRUMENT IN THE NHS?

Given the impressive potential value of 360 degree/MRF processes for supporting leadership development, some NHS organisations have used the method. The author has been involved in its use in several organisations, including the NHS, and some years ago conducted a study which compared the value of feedback to individuals of data from a Developmental Assessment Centre with those emerging from the 360 degree/MRF instrument she uses (Alimo-Metcalfe, 1990). The majority of individuals found the 360 degree/MRF data to be more detailed, of higher quality and of greater overall value than those received as part of the assessment centre method. However the author has become increasingly concerned that feedback on *competencies* might not offer specific feedback as to *the manner in which the competencies are utilised*. For example, a manager might be rated high on a competency relating to 'delegation', but she or he might perform the competency in a predominantly transactional or a transformational way.

Clearly competencies are crucial prerequisites for being perceived as an effective manager, but may tell us little or nothing about the manager's leadership style. If we focus on the former at the expense of the latter, we are in serious danger of reinforcing the current preoccupation in many organisations with transactional leadership. The question then arises as to how we might identify components for a 360 degree/MRF transformational leadership instrument.

What does transformational leadership look like in the NHS?

As was mentioned earlier in this chapter, most research on the nature of leadership has emerged from the USA. It cannot be presumed that notions of leadership are culture-free. Moreover these data may reflect, at least in part, bias by sector of employment and also by gender, since such research has been conducted on predominantly male samples. It was these concerns as to the possible lack of validity of such data that led to a major study being undertaken by the present author to investigate the nature of transformational leadership in the UK public sector.

Following her work with the Local Government Management Board (LGMB) and in particular with their national *Top Managers Programme*, the author was commissioned to investigate the nature of transformational leadership in local government. It was also hoped that she would conduct a parallel study in the NHS, which in fact was undertaken. A draft leadership questionnaire referred to as '*The Leadership Questionnaire*©' or *LQ*, was designed from constructs which emerged in interviews with 96 female and male top, senior and middle-level managers in local government and NHS organisations. Around 2000 constructs emerged and were grouped to form the basis of items (statements) to be included in the pilot *LQ*. Two slightly different versions of the *LQ* were created. They had 171 items in common but included some items relating specifically to the political context of the two types of organisation.

After piloting the questionnaires and seeking feedback from managers, the draft *LQ* was distributed to over 250 local government and NHS organisations. Around 30 questionnaires were sent to each organisation. Useable responses were received from around 1100 NHS managers and 1500 local government managers. Each group was factor analysed to establish the underlying structure or dimensions of transformational leadership. The results revealed a far more complex model of transformational leadership than that described in Bass's model (Bass, 1985). In addition to the four dimensions he identified, new factors emerged in the NHS sample. These related to openness to new ideas; honesty/tolerance regarding mistakes; being a visionary promoter of organisational achievements; intellectual flexibility; entrepreneurial risk taking; political skills; tenacity to achieve organisational goals; and integrity.

These findings, which emerge from an initial exploration of the data, suggest that Bass's model, while of relevance, is inadequate in describing how transformational leadership is viewed in the NHS. There are various possible reasons why the UK data reveal a more complex model. For example, this might have been due to the methodology employed to elicit the constructs, or to the fact that the original sample from whom the constructs were elicited included a large proportion of females; or it might have been due to the UK cultural context, or the fact that the data emerged from managers working in the complex environment of the public sector. These issues have yet to be addressed in future studies.

THE NHS TRANSFORMATIONAL LEADER

Nonetheless the new picture created of transformational leadership is one of an individual who is not only inspiring and visionary in his/her thinking, but is also genuinely interested in the concerns and aspirations of his/her staff as individuals, empowering them and creating opportunities for them to develop, and encouraging them to thinking critically and strategically. Such leaders celebrate and publicise their staffs' achievements, rather than 'owning the glory'. He or she is also very clear about what they want the organisation to achieve, inspiring others inside and outside the organisation to join them in this quest by involving them in developing the vision and the means to achieving it. On a personal level, they are honest, trusted for their integrity, open-minded, self-aware and tolerant of criticism and failures of others. While tenacious in achieving goals, they have a flexibility of approach which means that they will take risks and be entrepreneurial when attempting to achieve important outcomes.

What are the implications for the NHS of this new model of leadership?

If the NHS is to take these findings seriously, there are immediate implications for the development and selection of staff who have responsibilities for managing others. Selection and promotion criteria should be scrutinised to ensure that they not only include transactional managerial competencies, but also explicitly describe the transformational qualities and approaches.

The next challenge is to ensure that the techniques adopted for selection validly measure these behaviours/dimensions. This will require some creative thinking by those responsible for designing assessment processes since more of the commonly available techniques (including psychometric instruments, assessment centre activities and the rating scales adopted by assessors) are often geared primarily to measuring the transactional components of leadership. The manner, content and process of the selection interview must be thoughtfully designed to gather relevant data for judging transformational qualities and approaches. How will 'integrity', 'empowerment' or 'inspiring' be assessed? Will the selectors be chosen and trained in such a way that they can appreciate the difference between transactional and transformational leadership? The research findings which suggest that individuals tend to have a preference for a predominantly transactional or transformational approach (Bass, 1985) might suggest that there will be strong resistance to the selection of transformational individuals. After all, is it not highly likely that many senior managers have obtained their positions in the NHS precisely because they are primarily transactional in style? What are the chances of significant numbers of women (who, in general, are more likely to be predominantly transformational) being represented on senior or middle-level management appointment panels?

As exciting as the new form of leadership might appear to be, it will take time to create the necessary environment in NHS organisations for individuals who resemble this model to be appointed in significant numbers to senior and top levels. Can we break the mould? Apart from the selection process, the NHS needs to start now on approaching management and leadership development with this new model in mind. This is yet another major focus for attention.

Despite the above concerns, the author is hopeful that there will be influential individuals within the NHS who will recognise the importance of advocating and nurturing this new model of leadership. Perhaps the cost of inordinate stress, as revealed by the studies cited at the beginning of this chapter, will influence some movement in this direction. It is with this hope in mind that the author is developing a 360 degree/MRF version of the *LQ*.

CONCLUSION

The current emphasis of organisations, particularly those in the USA, to select and develop individuals with transformational leadership qualities is based on the consistent findings that staff who perceive their managers as predominantly transformational have significantly higher levels of job satisfaction, motivation, commitment and performance. Transformational leaders create transformational organisations and these are the ones, in both the public and the private sector, which are seen to be more effective in an environment of constant change.

The studies cited in this chapter provide guidance as to which competencies and qualities exemplify modern leadership in the NHS. Future studies should seek to evaluate whether the model of transformational leadership in the NHS presented here does affect not only individual, team and organisational performance, but also the psychological wellbeing of staff. While the former is more difficult to measure in health care organisations, the latter can be readily assessed and the findings from current research on stress levels in the NHS, and research on the effects of burnout, suggest that this is now an imperative.

References

Alimo-Metcalfe, B. (1990) 'An Evaluation of the Use of Synchrony™ in the Northern and Yorkshire Regional Health Authorities', Nuffield Institute Paper.

Alimo-Metcalfe, B. (1994) 'Gender bias in the selection and assessment of women in management', in M.J. Davidson and R.J. Burke (eds) *Women in Management: Current Research Issues* (London: Paul Chapman).

Alimo-Metcalfe, B. (1995) 'An investigation of female and male constructs of leadership and empowerment', *Women in Management Review*, 10(2), 3–8.

Alimo-Metcalfe, B. (1998) '360 degree feedback and leadership development', *International Journal of Selection and Assessment*, 6(1), 35–44.

Atwater, L.E. and F.J. Yammarino (1992) 'Does self–other agreement on leadership perceptions moderate the validity of leadership and performance predictions?' *Personnel Psychology*, 45, 141–64.

Barling, J. (1990) *Employment, Stress and Family Functioning* (Chichester: Wiley).

Bass, B.M. (1985) *Leadership and Performance Beyond Expectations* (New York: Free Press).

Bass, B.M. and B.J. Avolio (1984) *Improving Organisational Effectiveness Through Transformational Leadership* (London: Sage).

Bass, B.M. and F.J. Yammarino (1991) 'Congruence of self and others' leadership ratings of naval officers for understanding successful performance', *Applied Psychology: An International Review*, 40, 437–54.

Bass, B.M., B.J. Avolio and L. Atwater (1996) 'The transformational and transactional leadership of men and women', *Applied Psychology: An International Review*, 45(1), 5–34.

Boot, R., J. Lawrence and J. Morris (1994) *Managing the Unknown by Creating New Futures* (London: McGraw-Hill).

Borrill, C.S., T.D. Wall, M.A. West, G.E. Hardy, D.A. Shapiro, A. Carter, D.A. Golya and C.E. Haynes (1996) 'Mental Health of the Workforce in NHS Trusts', Institute of Work Psychology, University of Sheffield and Department of Psychology University of Leeds.

Brookings, J.B., B. Bolton, C.E. Brown and A. McEvoy (1985) 'Self-reported job burnout among female human service professionals', *Journal of Occupational Behaviour*, 6, 143–50.

Caplan, R.P. (1994) 'Stress, anxiety and depression in hospital consultants, general practitioners and senior health service managers', *BMJ*, 309, 1261–3.

Carrère, S., G.W. Evans, M.N. Palsane and M. Rivas (1991) 'Job strain and occupational stress among urban public transit operators', *Journal of Occupational Psychology*, 64, 305–16.

Charlton, J., S. Kelly, K. Dunnell, B. Evans and R. Jenkins (1993) 'Suicide Deaths in England and Wales: Trends in Factors Associated with Suicide Deaths', *Population Trends*, 69, 34–42.

Church A. (1997) Do you see what I see? An exploration of congruence in ratings from multiple perspectives', *Journal of Applied Social Psychology*, 27(11), 983–1020.

Cooper, C.L and J. Marshall (1976) 'Occupational Sources of Stress: A review of the literature relating to coronary heart disease and mental ill health', *Journal of Occupational Psychology*, 49, 11–28.

Firth-Cozens, J. (1994) 'Stress in Doctors: A Longitudinal Study', Report to Department of Health R&D Initiative on Mental Health of NHS Employees, University of Leeds.

Fletcher, C. (1993) *Appraisal: Routes to Improved Performance* (London: IPD).

Flocco, R. (1969) 'An examination of the leader behavior of school business adminstrators', *Dissertation Abstracts International*, 30, 84–85.

Golembiewski, R.T. and R.W. Boss (1992) 'Phases of burnout in diagnosis and intervention: Individual level of analysis in organization development and change', *Research in Organizational Change and Development*, 6, 115–52.

Harris, M. and J. Schaubroeck (1988) 'A meta-analysis of self–superior, self–peer and peer–supervisor ratings', *Personnel Psychology*, 41, 43–61.

Harrison, S., D.J. Hunter, G. Marnock and C. Pollit (1992) *Just Managing: Power and Culture in the NHS* (Basingstoke: Macmillan).

Heim, E. (1991) 'Job stressors and coping in health professions', *Psychotherapy and Psychosomatur*, 55, 90–99.

Hogan, R., G.J. Curphy and J. Hogan (1994) 'What we know about leadership', *American Psychologist*, June, 493–503.

Hough, L., M. Keyes and M. Dunnette (1983) 'An evaluation of three "alternative" selection procedures', *Personnel Psychology*, 36, 261–75.

Jackson, S.E., R.L. Schwab and R.S. Schuter (1986) 'Toward an understanding of the burnout phenomenon', *Journal of Applied Psychology*, 71, 630–40.

Karasek, R.A., D. Baker, F. Marxer, A. Ahlbom and T. Thorell (1981) 'Job decision latitude, job demands and cardiovascular disease: A prospective study of Swedish men', *American Journal of Public Health*, 71, 694–705.

Kelloway, E.K. and J. Barling (1991) 'Job characteristics, role stress and mental health', *Journal of Occupational Psychology*, 6(4), 291–304.

Lancet (1994) 'Burnished or burnt out: The delights and dangers of working in health', Editorial, 10 December, 1583–4.

London, M. and A.J. Wohlers (1991) 'Agreement between subordinate and self-ratings in upward feedback', *Personnel Psychology*, 44(2), 375–90.

London, M., A.J. Wohlers and P. Gallaher (1990) 'A feedback approach to management development', *Journal of Management Development*, 9(6), 17–31.

Maslach, C. and S.E. Jackson (1981) 'The measurement of experienced burnout', *Journal of Occupational Behaviour*, 2, 99–113.

McCall, M.W. and M.M. Lombardo (1983) 'Off the track: Why and how successful executives get derailed', Technical Report No. 21, Greensboro, NC, Center for Creative Leadership.

Moore, W. (1996) 'All stressed up and nowhere to go', *Health Service Journal*, 5 September, 22–5.

Morgan, G. (1993) *Imagination: The Art of Creative Management* (London: Sage).

Mount, M.K. (1984) 'Psychometric properties of subordinates' ratings of managerial performance', *Personnel Psychology*, 37, 687–702.

Offerman, L.R. and P.S. Hellman (1996) 'Leader behaviour and subordinate stress: A 360° view', *Journal of Occupational Health*, 1(4), 382–90.

Parker, D.F. and T.A. DeCottis (1983) 'Organizational determinants of job stress', *Organizational Behavior and Human Performance*, 32, 160–77.

Paton, C. (1996) *Health Policy and Management* (London: Chapman Hall).

Podsakoff, P. and D. Organ (1986) 'Self-reports in organisational research: Problems and prospects', *Journal of Management*, 12, 531–44.

Redman, T. and E. Snape (1992) 'Upward and Onward: Can staff appraise their managers?', *Personnel Review*, 21(7), 32–46.

Reilly, P.R., J.W. Smither and N.L. Vasilopoulos (1996) 'A longitudinal study of upward feedback', *Personnel Psychology*, 49, 599–612.

Rosener, J. (1990) 'Ways women lead', *Harvard Business Review*, Nov./Dec., 119–25.

Sargent, L.D. and D.J. Terry (1998) 'The effects of work control and job demands on employee adjustment and work performance', *Journal of Occupational and Organizational Psychology*, 71, 219–36.

Sparrow, J. and C. Rigg, (1993) 'Job analysis: Selecting for the masculine approach to management', *Selection & Development Review*, 9(2), 5–8.

Stewart, R. (1996) *Leading in the NHS: A Practical Guide*, 2nd edn (Basingstoke: Macmillan).

Sutherland, V.J. and C.L. Cooper (1988) 'Sources of work stress', in J.J. Harrell, L.R. Murphy, S.L. Sauter and C.L. Cooper (eds), *Occupational Stress: Issues and Development in Research* (New York: Taylor & Francis).

UNISON (1994) *Caring Against the Odds* (London: UNISON).

Wall, T.D., P.R. Jackson, S. Mullarkey and S.K. Parker (1996) 'The demands–control model of job strain: A more specific test', *Journal of Occupational and Organizational Psychology*, 69, 153–66.

Warr, P. (1990) 'The measurement of well-being and other aspects of mental health', *Journal of Occupational Psychology*, 63, 193–210.

Wexley, K.N., R.A. Alexander, J.P. Greenwalt and M.A. Crouch (1980) 'Attitudinal congruence and similarity as related to interpersonal evaluations in manager–subordinate dyads', *Academy of Management Journal*, 23, 320–30.

Wohlers, A.J. and M. London (1989) 'Ratings of managerial characteristics: Evaluation difficulty, co-worker agreement and self-awareness', *Personnel Psychology*, 42, 235–61.

11 The Influence of Middle Management upon Emergent Strategy: a case for more microempirical studies

Graeme Currie

INTRODUCTION

This empirical study takes a processualist approach (Mintzberg and Waters, 1985; Pettigrew *et al.*, 1992) to investigate the role of middle managers in the strategic change process in a UK NHS trust. The research focuses upon a specific aspect of strategic change, that of the emergence of a marketing 'strategy'. Such a strategy is seen as an exemplar for other change attempts in the UK NHS. This follows calls for more empirical studies of middle management (Dopson and Stewart, 1990). It also addresses the 'unjust' neglect of the public sector by academics carrying out research from a strategic management perspective (Ferlie, 1992; Lyles, 1990). The implications for a future research agenda are drawn out from the present study. These relate to the level of analysis of the strategic change process, conceptual frameworks brought to bear upon analysis, how theory and practice can be brought together and the preferred methodological approach in doing this. While the research site is one drawn from the UK, the analysis may well have resonance in other countries where the health service specifically and the public sector in general have also come under increasing pressure to change.

THE ROLE OF MIDDLE MANAGERS IN THE STRATEGIC CHANGE PROCESS

At least initially the Griffiths Report (DHSS 1983) appeared to strengthen the hand of middle management in the NHS (Pollitt *et al.*, 1991; Stewart and Walsh, 1992) in a way which flew in the face of contemporary trends in organisations to delayer middle management and add value in service organisations at the point of service delivery rather than the 'back office'. However, recently there appear attempts to marginalise and attack general managers in the NHS, which reflect more general organisational trends to restructure and 'thin out' layers of management (Cameron *et al.*, 1991; Cascio, 1993; Dopson *et al.*, 1992; Palmer, 1995). Within the NHS, for example, they have been attacked as the 'men in grey suits' and there have been purges of their numbers (for example, *Health Service Journal*, 1994a; 1994b; Hancock, 1994).

This begs the question of what is the appropriate balance between their elaboration and delayering, one which is notable by its absence in academic debate. As Dopson and Stewart (1990) state: 'If writing in this area [what is happening to middle management] is to amount to anything more than armchair theorising, it is crucial that more empirical work be done.' In the UK, NHS research remains focused upon top management in the organisation. For example, Ferlie *et al.* (1996, p. 24) concentrate their investigation at the level of the strategic apex of health trusts, yet they admit that this may be criticised because the focus is too high a level and that behaviour at the operational level may be unaffected by developments at this tier.

One of the main intentions of this chapter in setting out a future research agenda is to investigate the role of middle managers in the strategic change process. Is it the case that they are unimportant and that their delayering and marginalisation should continue? Alternatively have they an important role in strategic change which hitherto has been neglected? In this case a future research agenda would have a greater emphasis upon this group.

MARKET-DRIVEN CHANGE AND MARKETING

As a second strand in setting out a future research agenda, strategic change related to marketing is investigated as an illustration of the utility (or not) of a microempirical analysis of change which focuses upon process. The choice of issue reflects the status of the creation of the internal market and the subsequent market-driven change which followed as one of the dramatic features of NHS reforms over the past decade. For those unfamiliar with the UK NHS a coherent description and analysis of the impact of the internal market is provided by Ranade (1997). The important point to be made in the context of this chaper is that the attempts which followed legislation to orient middle managers towards the internal market provide an exemplar for efforts by top management to implement strategic change in general in NHS trusts.

As an exemplar for the study of strategic change, however, there remains a lack of research which explores the actual processes of market-driven change in the health care sector, apart from that of Whittington *et al.* (1994), who warned against the dangers of totalising accounts of change and identified a need for a more nuanced understanding of managerial control in relation to market-driven change in UK NHS hospital trusts. Importantly they pointed to the instability of new incentives in the way that they are appropriated for divergent, and unintended, purposes within the micropolitics of specific organisations and argued that there was a need for a detailed analysis of the processes of market-driven change. For example, the debate about the importation of generic marketing frameworks (Walsh, 1995; Wensley, 1990) which reflects a well-established debate generally about the importation of generic practices from private to public sector (Ackroyd *et al.*, 1989; Harrow and Willcocks, 1990; Ferlie *et al.*, 1996) is not based sufficiently on the micropolitics of the organisation. Therefore, from these studies, it is difficult to obtain a nuanced understanding of unintended consequences of market-driven change. This chapter seeks to examine the issue of the unintended consequences of market-driven change as well as those consequences intended by policy makers and top management and to assess the utility of a processual approach in achieving this.

THE CONCEPTUAL FRAMEWORK

In taking a processual view of strategic change, the emergence of strategic change is emphasised as an issue. Such a perspective allows a serious consideration of the role of middle managers in the strategic change process. The view taken of strategy is one which sees a process of strategic change rather than choice and of strategy as a 'pattern in a stream of actions' (Mintzberg and Waters, 1985). The strategic process is seen as one where there are patterns or consistencies realised despite, or in the absence of, intentions. Such a view questions the orthodox strategic change frameworks brought to bear by academics and practitioners which assume that 'top-down' rational planning or culture management is the appropriate way to go about changing the organisation.

Specifically analysis draws upon the framework developed by Pettigrew (1983; 1985) and applied to the UK NHS (Pettigrew *et al.*, 1992) which placed the concept of 'management of meaning' centrally. This refers to the process of symbol construction and value use designed to create legitimacy for one's own ideas, actions and demands and to delegitimate the demands of one's opponents. One of the most critical connections identified in the work of Pettigrew *et al.* is the way actors in the change process mobilise the contexts around them and in doing so provide legitimacy for change. In relation to the substantive issue under consideration, this chapter investigates the way key stakeholders advance different accounts of reorganisation towards a marketing orientation through the lens of legitimation and delegitimation. It assesses the extent to which top management can define meanings for middle management, and from this the circumstances under which middle management opposition is successful or unsuccessful are examined.

In applying such a contextualist approach with a middle management focus, this chapter investigates the interaction of top management intent ('view from the bridge') with the interest and activity of the middle management group ('view from the deck') in the creation of the realised marketing strategy. Thus a third strand of the future research agenda set out in this chapter investigates the usefulness of processualist views of strategic change as a sensitising device to inform academics and practitioners in their work.

METHODOLOGY

As a fourth strand of the future research agenda set out below, the question is posed regarding a preferred methodological approach to the study of strategic change. The study outlined here follows in the tradition of the ethnographic work analysing the impact of general management in the UK NHS of Strong and Robinson (1988). The usefulness of such an ethnographic approach to inform the work of academics and practitioners is considered. Therefore it is necessary to inform the reader of the details of the research design to facilitate consideration of its utility.

The research took place in a large acute hospital, Florence Hospital. Responsibility for the development of a marketing strategy, or a 'business development' strategy as it was labelled in the Florence Hospital, lay with the business development department, which was part of the central directorate. As far as setting the agenda for marketing activities was concerned, a focus upon the director of business development and business development manager with responsibility for marketing was relevant. Those who were viewed as recipients of the marketing strategy by the business development department, service managers and general managers within the clinical directorates are considered to be middle managers in this study. As an indicator of their role in the organisation, in the medical services directorate a service manager would take operational responsibility for 200 staff, including nurses and junior doctors, over five wards while the general manager to whom she reports would be concerned with the implementation of strategic change within the directorate.

Data were gathered in two stages over a period of two years (spring 1995 to summer 1997). Three cases of change in a hospital trust were examined, including business development policy and practices, via an iterative process of data gathering and analysis. In the first stage the researcher undertook general observation of the hospital environment; for example, via work shadowing, attendance at patient-focused care steering group meetings and attendance at management development workshops. Data were also gathered via informal interviews (25 in number) with executive directors (three), general managers (two), service managers (five) and other stakeholders in the process (for example, the organisation development manager and the patient-focused care manager). Notes were taken in these interviews.

In stage two, further observation was undertaken – for example, of marketing workshops – and a wealth of documentation was gathered. In particular, business plans provided a valuable insight to the strategic change process. A total of 31 formal interviews complemented the data gathered via documentation. These interviews were taped and transcribed. Five of the six general managers were interviewed, along with 11 service managers and five senior sisters. In stage two, formal interviews also took place with four executive directors (including the director of business development), the chief executive, the business development manager with responsibility for marketing, the external marketing consultant, the organisation development manager, the patient-focused care manager and the clinical effectiveness manager.

The iterative process of data gathering via observation, documentation and interviews with data analysis represented an inductive attempt by the researcher to generate 'thick description'. Following the final set of 31 interviews, themes were identified, some of which are reproduced in this chapter. These themes represent a second-order interpretation of organisational processes from the standpoint of the actors involved, collected and retold by the researcher, also representing a certain standpoint (Geertz, 1973).

A flavour of the data

The data presented here are necessarily constrained. However a flavour of the data is presented so that the reader can judge its richness. It contrasts the views of the two groups of stakeholders identified in this chapter in line with its emphasis upon strategic change as emergent. It focuses on the issue of the importation of generic marketing frameworks which is raised in the literature. In a fuller study empirical observations would be reproduced to a greater extent so that the reader could evaluate the analysts' rendition of events, thereby reducing the risk of bias from second-order readings. It is hoped that, in using such a methodology and conceptual framework in the chapter, an accessible, managerially relevant and authentic representation of the change process is provided. Such an approach draws its inspiration from academic work such as the account of change processes provided by Watson (1994).

VIEW FROM THE BRIDGE: SETTING THE AGENDA FOR MARKETING ACTIVITIES

The business development manager responsible for marketing across the trust, and the marketing consultant brought into support his work in particular, used generic marketing frameworks to problematise and provide solutions to environmental changes. They saw the important issue as one of cultural change:

> Marketing is a different way of looking at things. Instead of looking at it from what you do and your problems and being reactive, you have to be more proactive and look at how you are going to meet the needs of the market. (Marketing consultant)

> It is quite a task to change the culture of the organisation. You have got so many people who probably perceive marketing as a very commercially orientated function. Some people fail to see the relevance of it. It's the cultural issues which present trust managers with a lot of problems. For example, you could do a seminar about marketing to a group of Consultant Pathologists and be fairly safe, but with a group of Consultant Orthopaedic Surgeons you would be on very thin ice and would spend the first half of the meeting justifying the role of marketing in the health service. (Business development manager)

They stressed that the trust had to become more marketing oriented. In a unitarist view of the organisation they viewed any area resistant to a marketing orientation as a cultural lag. They viewed a marketing orientation as a superior view of the world, which was common sense within the environment faced by the trust. The business development manager and marketing consultant argued strongly that embracing a marketing orientation would solve directorate problems, even the problems of those clinical directorates claiming that their service activity was 'demand-led':

> The Florence is an emergency hospital therefore A&E [Accident and Emergency] is important. Embracing a marketing orientation solves the problems of team identity and poor presentation, both of which impact upon budget allocation. A change in culture via a 'marketing champion' [a new general manager] and customer care programmes, teambuilding, communications improvements encouraged a customer orientation.

VIEW FROM THE DECK: HOW THE MIDDLE MANAGERS FEEL ABOUT MARKETING ACTIVITIES

Many of the middle managers took a perspective in between the two extremes of a total rejection or acceptance of a marketing perspective.

They saw the success of the application of marketing as being dependent upon developing a robust and relevant understanding of marketing within the health service:

> I have come through the ranks of nursing. It's still a big part of me even though I'm a manager now. I'm uncomfortable with some of the changes; for instance, marketing. But if we don't go down that route another bigger provider from another city will undertake the surgical work and this will mean worse patient care for the people who live locally. (General manager: surgical services)

> I think that marketing per se as a function cannot be transposed into the NHS.... Because there are concepts and models that don't fit. Because we are not a profit-making organisation, we have ethical and clinical dimensions of selling and promoting our services. But there are some concepts that are applicable to the NHS such as looking at market segments, what position you are in the market, are you the market leader, are you just somebody who bubbles along sort of bread and butter stuff, and being more articulate in that sort of way. It focuses our decision making on whether we are going to increase contracts or not on a rational basis rather than a gut feel. (General manager: trauma and orthopaedics)

The emergence of a marketing orientation was particularly influenced by the sensitivity of the individual clinical directorates to the internal market, which was determined by the extent to which their activity was elective. Where activity was non-elective but was described as 'demand-led', this was used as an excuse to resist managerial initiatives, for example in the accident and emergency directorate.

Despite evidence of respondents' perceptions that the internal market lacked anchorage in 'reality' and resistance to the idea of marketing health care, there were activities which might be labelled 'marketing' going on in the trust. Often these represented unintended outcomes for the 'bridge' of the organisation, but ones which they retrospectively identified as 'marketing'. Many of the intended outcomes focused on building relationships with customers of service areas, although fragmentation was evident in who was perceived to be the customer for each clinical directorate. In some cases GPs were perceived as customers, while in other cases directorate managers within the same hospital were seen as customers:

> My particular experience [of setting the agenda] is actually being given the opportunity to go out and speak to the customers [GPs]. ... We set up a pain management programme and visited a lot of GPs who were all very keen. (Service manager: pain management)

> We have regular meetings with General Managers in other areas and what we are trying to do is relate the activity that we are doing and how we do it to their pressures, such as waiting lists. So, for example [I might say to them] I am funded to provide you with thirteen sessions per week but you are only using nine. Commercially that is costing you a lot of money. What do you want to do about it? Whereas before we just couldn't have those debates. We would have continued doing the same thing forever. (General manager: critical care).

There was evidence of the complexity of the health service in this latter statement. In common with many directorates, Critical Care perceived itself as different, as an area which needs to take less heed of generic marketing models. They defined the customer as other directorates and their marketing activity was focused on communicating with other middle managers in the trust.

DISCUSSION

In the same way that a flavour of the data was provided, analysis is also limited to the provision of a sufficient amount to allow the reader to assess the usefulness of the conceptual framework and methodology brought to bear. Drawing upon the work of Pettigrew *et al.* (1992) it can be expected that the interaction of the intent of those who seek to manage change and the interests and activities of middle managers results in a contested terrain. If this is the case then attempts to legitimate the concepts and perspective of marketing which the chief executive, business development manager and marketing consultant in particular support will also be subject to contestation. The contexts in which change operates are not inert or objective entities. Just as the chief executive, business development manager, marketing consultant and middle managers perceive and construct their own version of those contexts, so do they subjectively select their own versions of the environment around them, and seek to reorder the change agenda to meet perceived challenges and constraints.

Concentrating upon middle managers reveals scepticism from the middle managers towards marketing concepts. This was expressed as a feeling that marketing concepts had significant shortcomings when applied to health care activity in the Florence Hospital. They felt that the ethical and clinical dimensions of health care were not considered in generic marketing models. The advocates of a generic concept of marketing, the business development manager and marketing consultant, often found themselves in conflict with the efforts of other specialists who did not fully understand, share or endorse the marketing concept, its professed orientation or its view of the way organisations should be managed. In fact the middle managers were much more successful in mobilising inner and outer contexts to argue against a marketing orientation. Reservations about the applicability of marketing to the health service were linked back to the perceptions of the middle managers that the internal market was heavily regulated and that this made the introduction of marketing principles extremely difficult.

As far as the change in marketing orientation was concerned, there was not a clear picture. Views within the Florence appeared to be fragmented. Despite this, marketing activity went on and, while the middle managers perceived marketing as inappropriate because it aped private sector practices, some of the managers were taking on board those marketing concepts they saw as relevant. It appeared that generic marketing concepts and models were more likely to be well received if, for instance, they were linked to contextual characteristics.

If there was a marketing strategy it was emergent from the middle managers rather than a deliberate strategy with an emphasis upon a formal 'top-down' planning or cultural element, and was not labelled 'marketing'. It flowed from the characteristics of the outer and inner context which middle managers saw as legitimate: for example, the meeting of patient desires. The results of the emergent strategy were unpredictable, and the strategy of the chief executive and business development manager to gain a critical mass of support within the organisation for a marketing orientation was not successful.

The future research agenda

In addressing the future research agenda for organisational behaviour in health care, the importance of the middle managers in the UK NHS is emphasised. In the realisation of post-Griffiths reforms, the experiences of middle managers in the UK NHS are neglected in existing literature. Strategic change literature based on formal 'top-down' planning or culture management is limited in explaining change in the UK NHS because it does not take account of the role of middle managers beyond implementation of 'correct policies' determined at a higher level. The processualist literature (Mintzberg and Waters, 1985; Pettigrew *et al.*, 1992) allows us to gain a nuanced understanding of the change processes of the implementation of market-driven change and the realisation of a 'marketing strategy' which incorporates both intended and unintended consequences. It also allows us to address the strategic change issue in the NHS in general.

In the present case we see that the realisation of top management intent is an outcome of a process of political struggle and negotiation between the central directorate and middle managers. This chapter challenges pessimistic readings of the role of middle managers in the change process which suggest that they are implementors of change at best. That middle managers are important in the change process beyond a mere implementation role is a view shared in literature which questions the effectiveness of the orthodox formal planning models which dominate practitioners' and academics' thinking (Bower, 1970; Burgelman, 1983a; 1983b; Floyd and Wooldridge, 1992). In examining the emergence of a marketing strategy in the Florence Hospital, the role of middle managers is emphasised as both purveyors and recipients of change. In the case of marketing strategy, a critical element in market-driven change is decentralising responsibility to clinicians in middle management roles in the NHS (Dawson *et al.*, 1992). Market-driven change involves far more than simply grafting on new marketing activities. Spreading throughout the organisation, the process is often complex and slow (Whittington *et al.*, 1994). If this is the case, the present trend in the UK NHS to attack middle managers and delayer middle management in the interests of adding value to direct patient care may be misplaced. The question to be addressed in further research which pursues this point is the appropriate balance between the initial elaboration of managerial structures and their recent delayering.

For those concerned with managing strategic change, the case reinforces Pettigrew's view that it is necessary to gain legitimacy for a new perception of organisational performance and context so that contextually appropriate action follows. To understand this process there is a need for research to investigate the organisational context and the impact of the outer context of the organisation upon change processes. Factors inside and outside the organisation may set constraints on outcomes but, within these limits, the actions of individual actors have a part to play. Therefore any analysis must focus on the processes that are formed by these actions and, in particular, on the political nature of these processes. One of the critical issues for those concerned with strategic change is how and when to promote change, to reconcile the forces for stability and for change. In this, as Mintzberg (1990) suggests, it is a question of crafting strategy in the way a potter works at the wheel.

There may be those readers who claim to be more 'practically' focused, and argue that a research agenda which emphasises a microempirical analysis is less than useful or relevant. Certainly, in the course of this research, hospital staff found difficulty in recognising what the author was doing as research. They tended to refer to the study as 'whatever it is that you are doing, Graeme'. This lack of recognition reflects the emphasis in the health service upon 'hard' performance indicators and the use of quantitative data and administration of questionnaires to facilitate this. 'Hanging around and listening in', as the author resorted to describing what he was doing to hospital staff, was seen as exotic at best.

However this chapter argues strongly that such a microempirical approach is useful, relevant and even necessary. A quotation from the organisation development manager in Florence Hospital is illuminating in emphasising this:

> I would like to see much more deliberate and honest thinking at director level about the key things that are impacting upon and constraining this organisation. We need to take account of what different levels think, what the different realities in the organisation are. My intuition is that there are different views of reality and different levels in the organisation, all of which are valid. I don't think we take full account of those. We still think about the reality of senior managers and directors and impose it upon others. We need to think about the others' reality and try to persuade them of our reality.

An approach is required which gets close to the experiences of middle managers. Features of an ethnographic approach (Hammersley and Atkinson, 1995) need to be built into the research design: for example, case study work over a lengthy period, an emphasis upon process, data gathering via observation, qualitative data analysis and the production of rich description. While some may argue that it is inappropriate to generalise from one case, it can be argued that getting very close to managers in one organisation is a means of generalising about processes managers get involved in and about basic organisational activities rather than about 'all organisations' or 'all managers' (Watson, 1994). It is a matter of generalising theoretically rather than empirically, as Robert Yin puts it (Yin, 1984).

To summarise, there are two main issues for the future: firstly, we need to consider the response of our own community towards the type of research advocated in this chapter; secondly, alongside this, the response of host organisations which constitute the research sites is an important issue.

As far as the academic community is concerned, the research approach outlined is not favoured by some journals within the public sector field. Their preference lies with policy issues rather than analysis of managerial practice. Consequently they may not represent a fertile dissemination vehicle for those researchers carrying out microempirical analyses of change based upon case study work. Perhaps some academic journals need to address their assumptions about the validity of such 'coal face' work in the light of its usefulness which is claimed in this chapter.

In addition, host research sites for health services management research need to appreciate that a microempirical processual analysis can bring a nuanced understanding of the change process which is unavailable otherwise. Such research sites need to move away from their overreliance upon quantitative approaches to the evaluation of change. Academic researchers who seek to produce non-managerialist accounts of change are often marginalised in their work by practitioners. Yet the non-managerialist accounts remain managerially relevant. Health care organisations should engage with such research and seek to understand implications for themselves. It was rather disappointing in the case study that the researcher was never asked for feedback from the chief executive or other executive directors. The only organisational members who were interested in this were some of the middle managers who were fascinated by the authenticity of accounts produced. Many of the senior managers dismissed the researcher as 'academic' and therefore not relevant. The response to this may be that we should seek to make ourselves more relevant. However in this the danger is that we compromise our perspective and move towards senior managers' definitions of relevance, which remain managerialist.

Finally, in the case of both academic and practitioner communities, as is evident in the findings, middle managers, as a focus for research and as a resource for the organisation, are neglected. One reading of the future research agenda set out here is an emancipatory one, advocating the case for the middle managers to have a voice in research studies and in the strategic change process in organisations.

References

Ackroyd, S., J.A. Hughes and K. Soothill (1989) 'Public Sector Services and Their Management', *Journal of Management Studies*, 26(6), 603–19.

Bower, J.L. (1970) *Managing the Resource Allocation Process* (Cambridge, Mass.: Harvard University Press).

Burgelman, R.A. (1983a) 'A Model of the Interaction of Strategic Behaviour, Corporate Context and the Concept of Strategy', *Academy of Management Review*, 8(1), 61–70.

Burgelman, R.A. (1983b) 'A Process Model of Internal Corporate Venturing in a Diversified Major Firm', *Administrative Science Quarterly*, 28, 223–44.

Cameron, K.S., S.J. Freeman and A.K. Mishra (1991) 'Best Practices in White Collar Downsizing: Managing Contradictions', *Academy of Management Executive*, 5(3), 57–73.

Cascio, W.F. (1993) 'Downsizing: What Do We Know? What Have We Learned?', *Academy of Management Executive*, 7(1), 95–104.

Dawson, S., V. Mole, P. Winstanley and J. Sherval (1992) 'Management competition and professional practice', paper given at *Knowledge Workers in Contemporary Organisations Conference*, University of Lancaster, September.

DHSS (1983) *NHS Management Inquiry* (Griffiths Report) (London: HMSO).

Dopson, S. and R. Stewart (1990) 'What is Happening to Middle Management?', *British Journal of Management*, 1, 3–16.

Dopson, S., A. Risk and R. Stewart (1992) 'The Changing Role of Middle Managers in the United Kingdom', *International Studies of Management and Organisation*, 22(1), 40–53.

Ferlie, E. (1992) 'The Creation and Evolution of Quasi Markets in the Public Sector: A Problem for Strategic Management', *Strategic Management Journal*, 13, winter, 79–97.

Ferlie, E., L. Ashburner, L. Fitzgerald and A. Pettigrew (1996) *The New Public Management in Action* (Oxford: Oxford University Press).

Floyd, S.W. and B. Wooldridge (1992) 'Middle Management's Strategic Influence and Organisational Performance', *Journal of Management Studies*, 34(3), 465–85.

Geertz, C. (1973) *The Interpretation of Cultures* (New York: Basic Books).

Hammersley, M. and P. Atkinson (1995) *Ethnography: Principles in Practice* (London: Routledge).
Hancock, C. (1994) 'Managers Out for the Count', *Health Service Journal*, 105(5384) 17.
Harrow, J. and L. Willcocks (1990) 'Public Services Management: Activities, Initiatives and Limits to Learning', *Journal of Management Studies*, 27(3), 281–304.
Health Service Journal (1994a). 'Experts Warn Against Lack of Leadership in the NHS', 104(5427) 3.
Health Service Journal (1994b). 'The First Lesson for Margaret Beckett', 104(5427) 15.
Lyles, M. (1990) 'A Research Agenda for Strategic Management in the 1980s', *Journal of Management Studies*, 27(4), 363–75.
Mintzberg, H. (1990) *Mintzberg on Management* (London: Free Press).
Mintzberg, H. and J.A. Waters (1985) 'Of Strategies Deliberate and Emergent', *Strategic Management Journal*, 6, 257–72.
Palmer, C. (1995) 'White Collar, Black Hole', *The Observer Business*, 12 February, 6.
Pettigrew, A.M. (1983) 'Contextualist Research: A Natural Way to Link Theory and Practice', in E.E. Lawler (ed.), *Doing Research that is Useful in Theory and Practice* (San Francisco: Jossey Bass).
Pettigrew, A.M. (1985) *The Awakening Giant* (Oxford: Blackwell).
Pettigrew, A.M., E. Ferlie and L. McKee (1992) *Shaping Strategic Change* (London: Sage).
Pollitt, C., S. Harrison, D. Hunter and S. Marnoch (1991) 'General Management in the NHS: The Initial Impact 1983–1988', *Public Administration*, 69, 61–83.
Ranade, W. (1997) *A Future for the NHS? Health Care in the 1990s* (London: Longman).
Stewart, J. and K. Walsh (1992) 'Change in the Management of Public Services', *Public Administration*, 70, winter, 499–518.
Strong, P. and J. Robinson (1988) 'New Model Management: Griffiths and the NHS', Nursing Policy Studies Centre: University of Warwick.
Walsh, K. (1995) *Public Services and Market Mechanisms: Competition, Contracting and the New Public Management* (Basingstoke: Macmillan).
Watson, T.J. (1994) *In Search of Management: Culture, Chaos and Control in Managerial Work* (London: Routledge).
Wensley, R. (1990) 'The Voice of the Consumer? Speculations to the Limits of the Marketing Analogy', *European Journal of Marketing*, 24(7) 49–60.
Whittington, R., T. McNulty and R. Whipp (1994) 'Market-driven change in professional services: problems and processes', *Journal of Management Studies*, 31(6), 829–45.
Yin, R. (1984) *Case Study Research: Design and Methods* (Newbury Park: Sage).

12 MAPS for PAMS: managerial and professional solutions for professions allied to medicine

Annabelle L. Mark

The future for professions providing health care is uncertain, because of changes to the professional groups, their practices and the organisations in which they work. Much attention has been given to medicine in facing these dilemmas and considerable thought is given to nursing, not least because of the volume of staff involved, but further thought is required for professions allied to medicine (PAMS). This group is therefore the central focus of this chapter, which attempts to look across rather than just down through their relationships within the organisation of health care.

PAMS, like all professional groups, have both a national and an international context and both contexts now need to be considered, in relation to renegotiating professional roles, in order to arrive at appropriate managerial and professional solutions. This chapter identifies some of the current issues for this activity, for those working as therapists in health care, where the role is often misinterpreted as just the jam between the bread-and-butter roles of nursing and medicine. Solutions to the dilemmas that these professions face require analysis on a number of levels, but integration of the issues will be essential, if the future is to be understood and proactively managed to benefit patients and the professions alike.

The need to explore the future shape of organisations, and their relationship to and with therapists, is increasing as a range of options develop in the often fragmented and diverse provision of health care. Globalisation (Lessem, 1989; Schneider and Barsoux, 1997) of skills and services through an increasingly mobile workforce and the development of protocols for care, plus the potential emergence of multinational providers, may also soon be key factors in the availability of therapy services. At the same time the development of new professional agendas in health care is also likely to reshape organisations (JM Consulting, 1996; Schofield Report, 1996); but the assumptions upon which both organisational and professional developments are based need to be explored.

The organisation and professional development of therapy services, and PAMS in particular, must take account of a variety of factors internal and external to the health and social care agenda. Currently the agendas of organisations and the agendas of the professions increasingly diverge because of conflicts in objectives and perceptions of the present and future provision of health care. This situation is now further complicated in the UK by the proposals for collaborative working following the publication of *The New NHS – Modern Dependable* (Department of Health, 1997) which requires effective collaboration across sectors through new organisational structures. Existing health care organisations support the hierarchical relationships between health workers, often to the detriment of therapists (Department of Health, 1997a) and, despite a brief flirtation with markets, not only may such hierarchy be inappropriate for future care, but it may be counterproductive to innovation which is at the heart of professional practice and successful organisational development. Escaping such historically based hierarchies may be the only way to establish true team collaboration (Mark, 1998) across interprofessional groups, where professional rather than hierarchical relationships re-emerge to determine relationships. Hierarchies themselves become inappropriate as professional knowledge and expertise is increasingly dispersed (Mark and Elliott, 1998).

BACKGROUND

As a starting point for this analysis, questions arise about the future relevance of the medical model, which underpins health care and its organisation. This is most recently demonstrated in the development internationally of the evidence-based approach to medicine; it is in part a search for quantifiable outcomes which have been seen as also increasingly necessary to the organisation of health care (Lockett, 1997), yet on closer inspection the verisimilitude of either assumption is questionable and its achievement so far illusive. These perspectives may prove inappropriate in the longer term, as other more complex perspectives (Brown, 1995; Featherstone, 1995; Burrell, 1997) prove more convincing in understanding the 'webs of significance' (Geertz, 1973) which have been spun around health care.

A more appropriate approach may be to consider the drivers for change towards new organisational forms, which do not just take the organisation of health care, but the role of health care in its international, professional and cultural contexts, as a starting point (Bloor and Dawson, 1994). The organisational forms which will emerge from these interacting pressures may not fit the agendas of any of the current stakeholders, but will reveal what are the successful alliances. The identification of these may well be the key to the future, as the US Pew Commission Report on Revitalising the Health Professions for the 21st Century has pointed out: 'There will be major changes in the nature of all health professions with boundaries between professions becoming more flexible. Certain specific core skills will remain the sole province of one profession or speciality, but "shared skills" will be acquired by all health professionals'. Such shared skills may have cross-national, cross-educational and cross-professional implications.

Organisations are one of the drivers for change in the UK development of therapists or professions allied to medicine (PAMS), but are not so far perceived as a major factor in the development of this particular group of professionals. Contrary to this perspective, it is now being suggested that, rather than continuing to allow the existing organisational structures to drive the development of therapy services, therapists must control their future shape as a professional group, by developing and working in and through organisational structures appropriate to their purpose, or they will cease to maintain their distinctive professional roles.

Other imperatives such as the need for joint education and teamworking have been the subject of both national and international initiatives (Ovretvreit *et al.*, 1997; Soothill *et al.*, 1995) to improve the delivery of services and increase success in patient outcomes, but most assume the continued existence of national organisational structures. Organisational theory (Hassard, 1994; Chevalier and Segalla, 1996; Hatch, 1997; Burrell, 1997) however, and its varied applications to these disciplines, does have further contributions to make (Mark, 1994). The application of these theories may prove to be as significant in delivering objectives as the more specific desire for joint education and co-operation across professional boundaries. These boundaries which need to be crossed are described in relation to existing national structures, be they hierarchies or markets, but the emergence of organisational networks (Thompson et al., 1991) as an alternative form of working in health care (Kaluzny and Shortell, 1994; Ferlie and Pettigrew, 1996) is of growing significance. Defining networks continues to pose problems (Gummesson, 1994) and this is why commentators have concluded that understanding the goals of networks is the most appropriate way forward (Cravens et al., 1996). These goals are:

(a) to gain flexibility to cope with rapidly changing and intensively competitive markets;
(b) to develop the skills and resources needed to identify and quickly progress innovation;
(c) to achieve operating economies and efficiencies in order to offer value.

Networks, therefore, do not just involve building alliances, which in the context of markets operate anyway through the idea of relationship marketing (Cravens and Piercy, 1994) or in organisational hierarchies through the development of collaboration, as implied by the latest UK Government agenda (Department of Health, 1997). Networks involve the creation of alternative structures such as collaborative networks (Cravens and Piercy, 1994) to incorporate new agendas in therapy services, health care and the wider society. This approach, far from destroying effective team working, may enhance it.

UK CONTEXT

A number of local drivers for change are becoming apparent in the UK and now need to be recognised as such because they may also mirror an international pattern of change. The issue for professions, and the organisations in which they practise their skills, is to understand these pressures and develop strategies to enable desired outcomes to flourish by facilitating new *organisational* strategies and structures so that appropriate provision can be maintained. This provision must be appropriate to the customer and consumer, the professions and the organisations and societies in which they work. The organisations involved in health care in the UK, both in the public and the private sector, are proving less stable than the professional groupings they employ. This has implications for the way professionals relate to the control of their activities, where it is located and how credibility in professional practice is maintained (consumers must continue to trust the professional, whatever the shape or name of the organisation in which she or he works might be). For example, the loss of the public service ethic and its replacement by market values in the UK (Massey, 1993) had left many professions at odds with organisations which previously incorporated what they considered to be professional values within their own public service values (Ferlie *et al.*, 1996), although this may now be reversed by the latest reforms to the UK health system (Department of Health, 1997c).

In order to consider what the future options are for therapy organisations, there needs to be an understanding of what is meant by therapy services. In the UK a variety of definitions have existed, such as professions allied to medicine (PAMS) and professions supplementary to medicine (PSM), the latter being enshrined within the current legislation covering most of these disciplines; differences in definition are based largely on disputes over role boundaries (Stewart, 1990) either between PAMS (Øvretvreit, 1990) or between PAMS and others, notably doctors, nurses and social workers. Within this analysis the main groups included and referred to in general as PAMS are physiotherapy, occupational therapy, radiography, medical laboratory scientific officers, chiropody, dietetics, speech therapy, orthotists, orthoptists and also pharmacy and psychology. A number of other groups are also implied within governing legislation, but are not listed specifically here. Whichever groups are included, Professor MacLean (MacLean, 1997), chairing the review of the legislation in the UK, has noted that what they shared was more important than the way they differed, and thus they should be viewed collectively.

The legislation governing the majority of those working in these therapy areas in the UK is the 1960 Professions Supplementary to Medicine Act which has been the subject of an extensive review (JM Consulting, 1996). The results are to be incorporated in new legislation, the primary purpose of which is protecting the public. At the same time the legislation will need to be permissive rather than directive (McLean, 1997) enabling the inclusion of new groups and changing roles for those currently covered. Meanwhile others studying the area are incorporating an even wider definition to include nurses (McColl, 1997). This may be appropriate to the continuing professional developments within nursing exemplified by the role of the nurse practitioner and the interface with therapists, which differs in the UK from its US founding counterpart (see chapter 6) and is not yet governed as a professional role by statutory title. In considering this issue, it is appropriate to remember that the definitions, like the role boundaries, are fluid and inclusive, rather than exclusive, and reveal diverse and developing agendas in the provision of health care.

Organisationally what are the pressures for change which are now gathering pace? While these will be considered in a UK context, the international development of organisations, and therapy organisations in particular, is considered to be one of the key drivers of change, and will be included in the analysis.

DRIVERS OF CHANGE

In the UK the most prominent drivers of change at the present time are: political and economic, legislative, social, technological, organisational and professional.

Political and economic change is happening because of the election of a new government in the UK in May, 1997 after 18 years of the previous administration. This may affect some aspects of the health agenda; for example, while the volume of resources may not change dramatically, the focus and direction of them are changing. The removal of the internal market as a mechanism for allocative efficiency and its replacement with collaborative rather than competitive working redefines the agendas between purchasers of health care and providers in both the public and the private sector. However, organisationally, there may be problems as interorganisational collaboration means different things to different people (Mintzberg et al., 1996), and it is harder to get collaboration if there is a past history of independence or competitiveness to be overcome (Sheldon, 1979). Internationally the public/private sector mix in health care and the adoption of the new public management (Ferlie et al., 1996), where public money is used to buy care efficiently from either private sector organisations or public sector organisations, increasingly using private sector methods, are also influencing the scope and role of therapy care. In the USA for example, although the Clinton health reforms for greater state control failed to happen as originally planned

(Mechanic, 1995), the work of burgeoning therapy organisations (Quinn *et al.*, 1996) like Novacare continues to be significantly influenced by the state through the financial and outcome imperatives of Medicare and Medicaid programmes. This may prove, in the longer term, detrimental to innovative professional practice and continuing professional development. Likewise in the UK the failure of the health and social care systems, in which PAMS play a key role, to meet the needs of older people led to the setting up of a Royal Commission to investigate the issue (Department of Health, 1997b), and may reveal the issue to be as much one of organisational failures as of economic failures. Economic factors which are also influencing the agenda for the UK are those which relate to the European Union and the single currency. The mobility of the workforce, which is already a key issues for managers (Chevalier and Segalla, 1996; Esping-Andersen, 1996), will be facilitated by increasing political moves towards European union. Furthermore the European imperatives to keep public expenditure below a certain level in order to join in a single currency makes the continued outsourcing (Mark, 1994) or 'distancing' (Schofield Report, 1996) of services attractive to government, despite the political rhetoric to do otherwise.

The second driver for change is *legislative* – not just the definitions set out for legislation to regulate the health professions, (JM Consulting, 1996), or the terms and concepts implicit in the 1990 National Health Service and 1990 Community Care Act or the 1989 Children Act, but also those changes (HSJ, 1997a) adopted by the new administration. These will allow for the first time the employment of therapists by family doctors (GPs) in the UK incorporating all the economic advantages currently available to this group of therapists when working in the secondary care sector, in particular the transfer of pension rights of staff who choose in the future to work for GPs. It is hoped that this will help to change the working location of therapists away from hospital settings towards primary care settings (Sibbald, 1996), although this may be confounded by the imposition of poorly developed or untried organisational forms in primary care to further this objective. The new legislation to support developments for therapists recognises that the role of the therapy professions is becoming more autonomous, their activities more complex, their patients more vulnerable and their future more uncertain (Mark, 1997a).

A third driver for change which may need more consideration is the *social aspects* inherent in the changing role of women (Wilkinson and Howard, 1997), in terms of their place in the workforce (Esping-Andersen, 1996), their significant contribution as therapists in health and social care (Riska and Wegar, 1993; Halford and Leonard, 1997), and the flexible working patterns now required to maintain multiple roles as workers and carers/parents, plus a desire for professional self determination of both where and how they work. Women in general have shown an increased propensity to contribute to the development of successful small business start ups (*Barclays Review*, 1996) which facilitates independence, an advantage which the report says they value more than their male counterparts. This approach often secures a lifestyle which meets women's life cycle changes more appropriately than many large organisations can (Mark, 1994).

The *technological* developments in the delivery of appropriate care are a fourth significant factor, and perhaps more important to the shape and structure of therapy services are the developments in information management, which are facilitating both the delivery of care and the development of team practice in the UK (Morgan, 1993; Walby *et al.*, 1994). These factors which technologically shape services are exemplified by Novacare (Quinn *et al.*, 1996) in the USA, which has developed Novanet, an information management system, which captures and enhances system knowledge and professional development at the point of service delivery. A secondary development in technology, which will also affect the delivery of care, is the emergence of clinical guidelines or protocols for care (Garside, 1993; McColl, 1997). Such developments set out pathways of care through for example disease management (Hunter and Fairfield, 1997) which then become multinational treatments. This enables therapists to view the world rather than just their home country as their potential employer, in much the same way as doctors and nurses already do. Some countries and companies (Solovy, 1994) are already drawing on this perspective to widen job prospects and opportunities for qualified professionals, and this also challenges the manpower planning, recruitment and retention issues of individual countries.

Organisational developments which will drive change are both shaped by the political and economic agendas, which will see a reconfiguration of roles and relationships in the NHS in the UK. They will also be driven by the developments both national and international of other organisations which focus on providing various forms of therapy care. In the UK the dominant developments outside the statutory health sector have involved the following:

- unidisciplinary therapists, physiotherapists, occupational therapists, chiropodists and psychologists, for example, who wish to diversify into specific areas, much as care of the elderly or children, for which the statutory sector continues to experience staff shortages, (HSJ, 1997b);
- specific functions, such as the increasing need of professional advice to support litigation or claims against insurance or other third party payers, and more recently new government aims to assess need for the 'welfare to work' programme;
- alternative therapies, where therapists combine statutory therapies with alternative or complementary therapies, such as physiotherapy and osteopathy.

Currently therapists working in the UK outside the NHS and local authorities often work alone or in small groups, providing services contractually to the state sector; even without an internal market, this tendency is likely to continue and grow. Such a growing trend requires further analysis (Osborn and Hagedoorn, 1997) or it may prove to be counterproductive in the longer term if a critical mass of capabilities (Chesbrough and Teece, 1996) is required to maintain the provision and development of appropriate care for patients. However the state sector is not encouraging such developments as it looks for lowest-cost options vested in current practice and organisational structures which disempower certain professional groups. One solution to this problem, of developing organisations outwith the state sector, which enhances professional roles and practice, is to combine therapists within an inverted organisation (Quinn *et al.*, 1996), which is one where line becomes staff; (1) that is, the managers cease to manage, only providing a service to the professionals or field experts, who thus control the organisation and determine things such as organisational rewards. Such structures can be seen, not only as a way for therapists to reassert control over their activities, but also as providing:

(a) reductions in direct costs;
(b) flexibility in coping with periods of change;
(c) release of management to focus on key organisational tasks;
(d) access to expertise not available in-house (Mark, 1994).

The drivers for this type of organisation in the wider context have been encapsulated by organisational theorists in notions of outsourcing and knowledge-based networking (Quinn, 1992), not least because as Quinn *et al.* (1996) suggest, 'the capacity to manage human intellect and to convert it into useful products and services is fast becoming the critical executive skill of the age'. Such alliances facilitate gains in technical capability, tacit knowledge (Polanyi, 1962) and an understanding of rapidly changing markets faster since they may not require individuals or units to unlearn (Lyles *et al.*, 1996) traditional routines (Osborn and Hagedoorn, 1997).

Organisations have now caught up with the professions (Macdonald, 1995) in valuing knowledge as the most important commodity at its disposal. What must therefore be viewed as significant is the degree to which, meanwhile, public sector organisations in the UK, which are highly professionalised (Hood, 1991; Willcocks and Harrow, 1992), have been seeking through various devices to corporatise professional activity (Hunter, 1994). This is done by giving doctors 'legitimate' or 'position' power (French and Raven, 1958) in other words hierarchical authority over others in the organisation, using for example the clinical directorate model and latterly the development of clinical governance (Department of Health, 1998), with the doctors also retaining the 'expert' power which they already enjoy (Mark, 1997b).

The need for therapists in the UK to take a broader view of these issues is now arising because of the dual frustrations for them of either remaining within the NHS, with the constraints and problems which that poses, or working outside it largely as lone operators or as small partnership, which causes problems from both a professional and an administrative standpoint. The third option which now presents itself, of working within small to medium-sized organisational teams associated with family medicine led by GPs, will therefore have many attractions. It may also have the disadvantages and risks associated with the clinical directorates in NHS trusts, where doctors are once again in the driving seat to manage and define roles (Mark, 1997b). Furthermore, because of the unique employment position that general practitioners enjoy as independent contractors to the NHS, they would also become employers of the therapists who choose to work in general practice, thus exercising an even greater degree of control than their clinical directorate counterparts in the secondary care sector.

Professional developments driving the change are embodied in the changing legislation in the UK which, by moving towards protection of title, is advancing the traditional view of professionalisation (Turner, 1995) as a strategy for achieving occupational monopoly. Yet the proposed legislation in the UK also implies future inclusion rather than exclusion of other groups developing appropriate therapies if they fit the criteria of involving:

(a) invasive procedures or clinical intervention with the potential for harm;
(b) the exercise of judgement by the unsupervised professional which can have a substantial impact on patient health or welfare (JM Consulting, 1996).

So applications from tattooists, beauty therapists or those providing colonic irrigation cannot be ruled out. More seriously, however, the new legislation does provide the opportunity for complementary therapies to gain statutory recognition which, given the dual activities of, for example, homeopathy and pharmacy, hypnotherapy and psychology and even some beauty therapies and the developing role of general practice, will provide some safeguards and reassurances for the consumer. Furthermore the legislation may also enable the notion of generic care at a professional level (Schofield Report, 1996) to be incorporated in the workforce if boundaries become really flexible, as state organisations in particular wish them to do, and if the nursing profession continues to reject any involvement with the generic workforce, as it did at the 1998 Royal College of Nursing conference.

The continuing tensions inherent in the development of the therapy professions in the UK have been exemplified by the argument about which titles to protect, the generic one of health professional, as suggested by the College of Radiographers (Smith, 1997) or one particular to the occupations involved, such as physiotherapy or radiography. Other professionalised groups, now also the subject of legislative reviews (JM Consulting, 1998), incorporate a number of occupations, including nurse, midwife and health visitor, or physician, general practitioner and surgeon, but generic titles of nurse or doctor are the common usage for consumers. However, within the Medical Act 1983, section 49, which governs all medical practitioner titles, it is interesting to note that the term 'apothecary' remains a protected title, possibly because of its wider international meaning, yet in the UK it seems a somewhat archaic retention by medicine of activities now undertaken by pharmacists, who, in changing the title but not the activity, have removed it from statutory control by the medical profession alone. Thus protection of title, although seemingly important, gives no guarantees about the future for either the status of the title or the activities linked to it. There is, however, in the wider social context, some simplicity for the consumer in proposing a future which in the UK at least includes the shared but differentiated concepts of medical practitioner, health practitioner and nurse practitioner. The last of these is as yet not only an unprotected title but also an ill-defined role in the UK (see Chapter 6), although its future may be as the apex of clinical nursing, if nurses rather than doctors are allowed to define it. Such a solution fits with the three pay spines proposed for the future workforce (NHSE 99) and also leaves arguments about the nature of professionalism to those others who wish to consider issues wider than consumer protection, which is the main concern of the legislation and the public it is designed to protect. The defining link between these three titles for consumers is clarity and simplicity, as each would indicate a range of skills and professional judgements practised under the control of one professional body, the General Medical Council GMC for doctors, the United Kingdom Central Council UKCC for Nurses, Midwives and Health Visitors and a third body for therapists, perhaps based on the Council for the Professions Supplementary to Medicine, to which the professions are answerable and the consumer can appeal for either information or protection. The three groups would encapsulate all health workers giving professional services and provide role clarity for the consumer even within the context of a single legislative framework now being proposed for all health professionals (JM Consulting, 1998).

FUTURE DIRECTIONS

The issues addressed so far indicate the complexity of the pressures for change but are by no means exclusive. What is important is determining how to proceed within this environment, using where appropriate some of the more theoretical perspectives which can make some sense of these issues and supply some ways of proceeding in the future. A number of issues are now considered in this context. The development in professional status can be said to mirror the four levels of intellect operating in an organisation (Quinn *et al.*, 1996) in the following way:

- know what equals knowledge;
- know how equals application;
- know why equals cause and effect and intuition;
- care why equals motivation, adaptation and innovation.

Innovation is therefore the pinnacle to which professionals lay claim using their unique skills and independent judgement without being subject to control by others, and it is in the need for innovation that issues relating to the future boundaries between roles and their development reveal tensions with others, and have a bearing on the desire to improve teamworking (Øvretvreit *et al.*, 1997; Soothill *et al.*, 1995).

Innovative work, which is required because of all the previously mentioned drivers of change, could require different approaches for the future, as the environmental and organisational contexts are no longer appropriate. One such development in organisational terms is the notion of the 'virtual organisation'. Virtual organisations subcontract anything and everything because this maximises the potential for flexibility and innovation (Chesbrough and Teece, 1996). However the virtue of virtuality will depend on whether the innovation required is *autonomous* (it can be pursued independently of other innovation) or *systemic* (the benefits can only be realised in conjunction with complementary innovations by others) (ibid., 1996). Innovation in therapy services falls into the category of systemic rather than autonomous activity; that is, it requires knowledge and assistance from others to be successful. While independent practice has many practical attractions for therapists who wish to determine their own future, the need to sustain professional competence to allow innovation to happen, which in turn sustains professional credibility and contracts for work, requires a more sophisticated organisation than most current alternatives to the statutory sector in the UK. The increasing provision of care in patients' homes and other non-statutory environments lends itself to virtuality. However, some concerns about virtuality making management of government and its agencies impossible (Mintzberg, 1996), unless normative or professional values and beliefs alone take control again, fail to take account of the integrative role of consumers (see Chapter 5) in such arrangements, which could prove more important than is currently assumed.

Inverted (Quinn *et al.*, 1996) virtual (Chesbrough and Teece, 1996) networks (Cravens *et al.*, 1996), would mean professionals are responsible for the provision and development of the professional resource working in innovative teams in a contractual relationship with providers/commissioners of health provision across organisational and, possibly, national boundaries. They would allow professionals a critical mass of professional and support capability, allowing some *freedom from bureaucratic process*, because if the organisation is big enough it will be provided by support staff, *freedom to innovate and develop practice* with other peer professionals, and *freedom from role definition and boundary control by other professional groups*. The first freedom is provided in the UK by remaining in the statutory sector at either a primary or secondary care level; the second freedom often still requires changes in attitude within and between professionals (Mark, 1998) in both statutory and independent sectors; the third freedom requires the negotiation of new relationships for which separation from professional hierarchies within existing organisations may be the only way to ensure that these new relationships are negotiated, rather than imposed by others. Outsourcing activities in this way or developing what Cravens *et al.* (1996) have described as a 'transactional networks', as opposed to a collaborative network, the former focusing on the exchange relationships which link organisations while the latter focuses on common characteristics, is thus not only a way of ensuring that the lid is kept on public expenditure levels, which the political and economic agenda suggests might be desirable; it also provides a method to renegotiate roles outside the hierarchical traps of existing UK health organisations. This is perhaps also why, in bringing together a critical mass of capabilities in this way through an inverted organisation, Novacare is one of the fastest growing health companies in the USA, operating in over 40 states (Quinn *et al.*, 1996).

Furthermore, by operating as a third party in the interorganisational relationships between health and social care policies, this model of virtual network therapy (Cravens *et al.*, 1996) may also provide some dynamics to facilitate developments between the two agencies. Such alternatives also give some answers to the tensions in this transformation of professionals identified by Broadbent *et al.* (1997) as:

(a) a need to accommodate professional autonomy compared to an organisational need to develop strategic control;

(b) the development of a professional identity rather than simply the development of an organisational identity;
(c) a respect for professional practice and the need to ensure change in that practice.

Matching organisation to innovation or making the choices about being virtual depend on the capabilities which are needed and the type of innovation, systemic or autonomous as indicated in Figure 12.1. Resolving this dilemma in the UK is far more complex than Chesbrough and Teece (1996) suggest. However what is important is the degree to which the capabilities needed already exist outside the health and local authority sectors, both now and in the future.

Regarding the need to create them, the question which arises is, given the rapid and continuing level of change in health and social care organisations exemplified in the UK, compared to the relative stability (Bloor and Dawson, 1994) of professional groups such as the

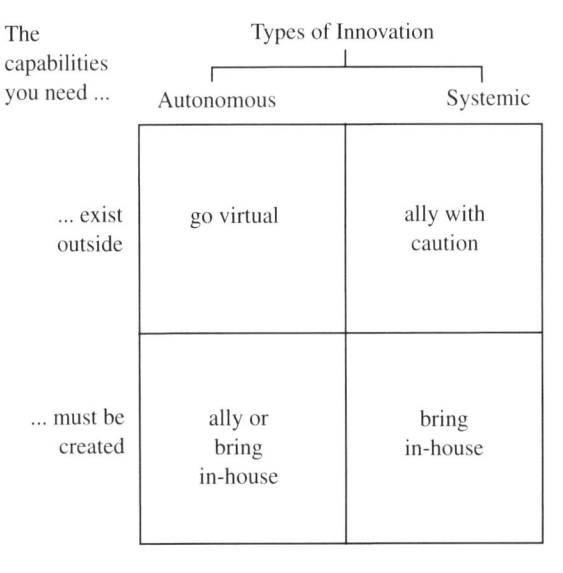

The capabilities you need ...	Types of Innovation	
	Autonomous	Systemic
... exist outside	go virtual	ally with caution
... must be created	ally or bring in-house	bring in-house

Figure 12.1 Matching organisation to innovation
Reprinted by permission of *Harvard Business Review* from 'When is Virtual Virtuous? Organizing for Innovation' by Henry W. Chesbrough and David J. Teece, January–February 1996. Copyright 1996 by President and Fellows of Harvard College; all rights reserved.

therapists, where are such innovative practices likely to be systematically created and sustained? Continuity required for systematic innovation in therapies may be best located within professional groups where some continuity exists. This may then lead to professional working in groups practising on behalf of patients via contracts with state providers of care. Such arguments for alternatives to direct state provision are further reinforced by the lack of NHS leadership, in providing direction for PAMS. This may also ultimately mean that the state sector is left only with direct control of generic workers who, contrary to some suggestions (Schofield Report, 1996), may not be professionalised, especially if existing professional groups continue to reject control of them, because they threaten the heart of professional practice.

Furthermore if innovation, as McLean's (1997) proposals for PAMS legislation in the UK suggest, is going to enable changing roles and boundaries between, for example, complementary and statutory therapies to develop, it may be that such innovation can only occur, with any rapidity, outside the public sector in organisations where therapists are in charge of their destinies so that professional rather than organisational controls provide consumer protection. This is also why the grey areas of the boundaries between state and private provision, which are issues of contention in the UK, may be as nothing compared to the changing boundaries between national and international providers. Boundaries and relationships become increasingly complex whether focused on the activities (for example, at least 20 different professions are now involved in the provision of ultrasound in health care) or on organisations, within the public/private or national/international axis.

In conclusion the complex array of drivers of change in the delivery of therapy services need new perspectives on future organisations which can most effectively deliver care. Inverted virtual networks are one possible solution, but would require acceptance by statutory and private sectors as well as the therapy professions themselves. If they do not consider such options therapists may continue to be frustrated in developing and defining their roles to deliver effective care into the next millennium.

Note

1. Line relationships indicate the vertical flow of authority and control in an organisation, while staff relationships are appointed as extensions of their

superiors and carry no authority except of an informal kind as personal assistant on policy adviser.

References

Barclays Review (1996) *Women in Business* (London: Barclays Bank PLC Small Business Service).
Bloor, G. and P. Dawson (1994) 'Understanding Professional Culture in Organisational Context', *Organisation Studies*, 15(2), 275–95.
Broadbent, J., M. Deitrich and J. Roberts (eds) (1997) *The End of the Professions* (London: Routledge).
Brown, A. (1995) *Organisational Culture* (London: Pitman).
Burrell, G. (1997) *Pandemonium – towards a retro organisation theory* (London: Sage).
Chevalier, F. and M. Segalla (1996) *Organisational Behaviour and Change in Europe. Case studies* (London: Sage).
Chesbrough, H.W. and D.J. Teece (1996) 'When is Virtual Virtuous? Organizing for Innovation', *Harvard Business Review*, 74(1), 65–73.
Cravens, D.W. and N.F. Piercy (1994) 'Relationship marketing and collaborative networks in service', *International Journal of Service Industry Management*, 5(5), 39–54.
Cravens, D.W., N.F. Piercy and S.H. Shipp (1996) 'New Organisational forms for competing in highly dynamic environments: the network paradigm', *British Journal of Management*, 7, 203–18.
Department of Health (1997a) *Getting involved and making a difference – purchasing and the professions allied to medicine* (Leeds: Department of Health).
Department of Health (1997b) Press Release 1997/379, 'Royal Commission on Long Term Care for the Elderly'.
Department of Health (1997c) *The New NHS – Modern Dependable* (London: HMSO).
Department of Health (1998) *A First Class Service – quality in the new NHS* (London: HMSO).
Esping-Andersen, G. (ed.) (1996) *Welfare States in Transition* (London: Sage, in association with the United Nations Research Institute for Social Development).
Featherstone, M. (1995) *Undoing Culture – globalisation, post modernism and identity* (London: Sage).
Ferlie, E.B. and A.M. Pettigrew (1996) 'Managing through networks: some issues and implications for the NHS', *British Journal of Management*, 7, special issue, March, 81–99.
Ferlie, E.B., L. Ashburner, L. Fitzgerald and A. Pettigrew (1996) *The New Public Management in Action* (Oxford: Oxford University Press).
French, J. and B. Raven (1958) 'The bases of social power', in D. Cartwright (ed.), *Studies in Social Power* (Ann Arbor: Michigan Institute for Social Research).
Garside, P. (1993) *Patient-focused care: a review of seven sites in England* (Leeds: NHSME).
Geertz, C. (1973) *The Interpretations of Cultures* (New York: Basic Books).

Gummesson, E. (1994) 'Service Management: an evaluation and the future', *International Journal of Service Industry Management*, 5(1), 77–99.

Halford, S. and P. Leonard (1997) *'Gender and Identity in National Health Service Organisations'*, UK ESRC Research project in progress, University of Southampton.

Hassard, J. (1994) 'Post-modern organisational analysis: towards a conceptual framework', *Journal of Management Studies*, 31(3), 303–24.

Hatch, M.J. (1997) *Organisation Theory – modern, symbolic and post-modern perspectives* (Oxford: Oxford University Press).

Hood, C. (1991) 'A public management for all seasons?', *Public Administration*, 69(4), 3–19.

HSJ (1997a) 'News-Government allows all practice staff to join NHS pension plan', *Health Services Journal*, 19 June, 107(5558) 7.

HSJ (1997b) 'News – NHS faces pharmacist shortage', *Health Services Journal*, 26 June, 107(5559) 10.

Hunter, D. (1994) 'From Tribalism to Corporatism: the managerial challenge to medical dominance', in J. Gabe, D. Kelleher and G. Williams (eds), *Challenging Medicine* (London: Routledge).

Hunter, D.J. and G. Fairfield (1997) 'Managed Care: Disease Management', *British Medical Journal*, 5 July, 315(7099) 50–3.

JM Consulting (1996) 'The regulation of health professions – report of a review of the Professions Supplementary to Medicines Act 1960 with recommendations for new legislation', report for UK Health Department, Leeds.

JM Consulting (1998) 'The regulation of nurses, midwives and health visitors – invitation to comment on issues raised by a review of the Nurses, Midwives and Health Visitors Act 1997', JM Consulting, Bristol.

Kaluzny, A. and S. Shortell (1994) 'Creating and managing the future', in S.M. Shortell and A. Kaluzny (eds), *Health Care Management, organisation design and behaviour* (Albany, NY: Delmar Publishers).

Lessem, R. (1989) *Global Management Principles* (Hemel Hempstead: Prentice-Hall).

Lockett, T. (1997) *Evidence-based and Cost-effective Medicine for the Uninitiated* (Oxford: Radcliffe Medical Press).

Lyles, M., G. von Krogh, J. Roos and D. Kleine (1996) 'The Impact of Individual and Organisational Learning on Formation and Management of Organisational Co-operation', in G. von Krogh and J. Roos (eds), *Managing Knowledge* (London: Sage).

Macdonald, K.M. (1995) *The sociology of the professions* (London: Sage).

Mark, A. (1994) 'Outsourcing Therapy Services', *Health Manpower Management*, 20(2), 37–40.

Mark, A. (1997a) Presentation on policy and legislation to the Annual Conference of Occupational Therapy Managers in York, entitled 'Managing the opportunities of change', organised by the College of Occupational Therapists, London.

Mark, A. (1997b) 'Doctors in Management', in P. Anand and A. McGuire (eds), *Changes in Health Care* (London: Macmillan).

Mark, A. (1998) 'Team working in primary and secondary care – identifying development dilemmas', *Clinician in Management*, 7(1), 46–8.

Mark, A. and R. Elliott (1998) 'Consuming the market-context, subtext, pretext and dialogue in the UK National Health Service', paper presented at the 3rd International Conference on Organisational Discourse, Kings College London, July.

Massey, A. (1993) *Managing the Public Sector – a comparative analysis of the United Kingdom and the United States* (Aldershot: Edward Elgar).

McLean, S. (1997) 'Steering through uncharted waters', paper presented at The Regulation of Health Professions Conference, Royal College of Surgeons, London, January.

McColl, E. (1997) 'Protocol for the systematic review of the effectiveness of guidelines in professions allied to medicine', University of Newcastle, Centre for Health Services Research.

Mechanic, D. (1995) 'Failure of Health Care Reform in the USA', *Journal of Health Service Research and Policy*, pre-launch issue, October.

Mintzberg, H. (1996) 'Managing Government Governing Management', *Harvard Business Review*, 74(3) May/June, 75–80.

Mintzberg, H., J. Jorgensen, D. Dougherty and F. Westley (1996) 'Some surprising things about collaboration: how people connect makes it work better', *Organisational Dynamics*, spring, 60–70.

Morgan, G. (1993) 'The implications of Patient-Focused Care', *Nursing Standard*, 7(52), 15 September, 37.

NIISE (1999) *Briefing Note BNO5: agenda for change–modernising the NHS pay system*. (Leeds: DOH).

Osborn, R.N. and J. Hagedoorn (1997) 'The institutionalisation and evolutionary dynamics of interorganisational alliances and networks', *Academy of Management Journal*, 40(2), 261–78.

Øvretveit, J. (1990) *Therapy Services organisation, management & autonomy* (Reading: Harwood Academic Publishers).

Øvretveit, J., P. Mathias and T. Thompson (eds) (1997) *Interprofessional Working for Health and Social Care* (Basingstoke: Macmillan).

Polanyi, M. (1962) *Personal Knowledge: towards a Post Critical Philosophy*. (London: Routledge).

Quinn, J.B. (1992) *Intelligent Enterprise* (New York: Free Press).

Quinn, J.B., P. Anderson and S. Finkelstein (1996) 'Managing Professional Intellect – making the most of the best', *Harvard Business Review*, 74(2) March/April, 71–80.

Riska, E. and K. Wegar (1993) *Gender, Work & Medicine: women and the medical division of labour* (London: Sage).

Schneider, S.C. and J.-L. Barsoux (1997) *Managing Across Cultures* (Hemel Hempstead: Prentice-Hall Europe).

Schofield Report (1996) 'The future healthcare workforce – the steering group report', Health Services Management Unit-Project Steering Group, University of Manchester.

Sheldon, A. (1979) *Managing Change and Collaboration in the Health System: the paradigm approach* (Berlin: Oelgeschlager, Gunn & Hain).

Sibbald, B. (1996) 'What is the future for a primary care led NHS?', *National Primary Care Research and Development Centre Series* (Oxford: Radcliffe Medical Press).

Smith, P. (1997) 'A Radiographer's Perspective', paper presented at The Regulation of the Health Professions Conference, Royal College of Surgeons, London, January.

Solovy, A. (1994) 'New Power Strategies (the battle for control)', *Hospital & Health Networks*, 68(24), 24–34.

Soothill, K., L. Mackay and C. Webb (1995) *Interprofessional Relations in Health Care* (London: Edward Arnold).

Stewart, R. (1990) *Developing Role Theory for Research in Management* (London: Pan Business Books).

Thompson, G., J. Frances, R. Levacic and J. Mitchell (1991) *Markets, Hierarchies and Networks – the co-ordination of social life* (London: Sage, in association with the Open University).

Turner, B.S. (1995) *Medical Power and Social Knowledge*, 2nd edn (London: Sage).

Walby, S., J. Greenwell, L. Mackay and K. Soothill (1994) *Medicine & Nursing – professions in a changing health service* (London: Sage).

Willcocks, L. and J. Harrow (eds) (1992) *Rediscovering Public Services Management* (London: McGraw-Hill).

Wilkinson, H. and M. Howard (1997) *Tomorrow's Women* (London: Demos).

13 Evidence into Practice? An exploratory analysis of the interpretation of evidence

Louise Fitzgerald, Ewan Ferlie, Martin Wood and Chris Hawkins

INTRODUCTION

This chapter draws on a pair of research projects, using similar methodologies which seek to explore the processes by which innovations diffuse into clinical practice. The research sets out to understand, inform and improve the processes of evidence-based medicine. Evidence-based medicine (EBM) involves the diffusion of evidence, particularly new or updated evidence, into clinical practice. As such, it includes complex processes of understanding, deciding, evaluating, communicating and agreeing. At the outset one would also stress that EBM involves change and change processes. We hope this briefly emphasises the first tenet of this chapter, that we are investigating and describing complex processes. This chapter will focus on one problematic aspect of EBM which, with a few notable exceptions (Williamson, 1992; Dawson, 1995; Berg, 1997), has been particularly neglected: the nature of 'the evidence' itself. Do we understand what is meant by scientific evidence? Closely associated with the nature of the evidence are the meanings ascribed to it by professionals. What are their perceptions of evidence? Are they uniform? Finally, in considering the diffusion of evidence and its use in practice, what are the attributes of evidence which make it credible and, therefore, potentially, used?

In considering how evidence is conceived within the processes of diffusion, one can detect a variety of models in use. Williamson (1992) usefully details the 'knowledge-driven' and 'problem-solving' models of knowledge production and use. The former model is based on the assumption that the sheer fact that knowledge exists presses towards its use. The latter model suggests that research provides evidence and conclusions that help to solve problems. These models are closely akin to those described in the private sector literature as 'push' and 'pull' models of innovation and diffusion. Continuing his critique, Williamson points out that the fundamentally different purposes of research and policy making mean that research is unlikely to flow into action. Research is concerned with extending and questioning the boundaries of the known, whilst policy making utilises present knowledge and 'concretises' it. Deykin and Haines (1996), in an illuminating review of research on the diffusion of medical innovations, draw attention to the oversimplistic nature of many models. They underline the need for more complex treatments and, drawing on Rogers (1983), propose a five-stage model of innovation and diffusion. While this model is more sophisticated, it remains a linear model, in which 'knowledge' appears unproblematic. The authors refer briefly to research from other sectors, outside health care, without drawing heavily on it in their conclusions.

In many publications, it is evident that this linear model remains the conceptual model underpinning much thinking. For example, there is work which focuses on the need for better information and the improved 'flow' of information (Weed, 1997). Similarly the production and use of clinical guidelines has been the subject of considerable discussion (Field and Lohr, 1990; Eddy, 1982; 1990; Grol *et al.*, 1993). Advocates argue that protocols will render medical practice more scientific and reduce variations. Rappolt (1997) shows that the extensive development of clinical guidelines and protocols in Canada has had limited effects on clinical practice. Mulrow (1996) likewise provides a balanced review of the strengths and weaknesses of protocols and guidelines. Underlying many of the 'problems' of guidelines, which she highlights, is the lack of any linearity in the reality of knowledge production and diffusion. More importantly many of the 'problems' of guidelines raise issues concerning the credibility and contestablility of the evidence itself.

Much useful knowledge might be derived from research in the private sector and particularly from models of technology transfer. A similar range of models exist here. For example, Williams and Gibson (1990) show how different phases of research were based on four different models of diffusion. The appropriability model is close to the 'push' model described above; this model suggests that, if one has sound scientific ideas, rigour in the research approach and good communication, then the technology will transfer. The dissemination model argues that the critical success factors include good science, but places greater stress on strong networks and supportive Human Resource Management (HRM) policies. The next model is labelled the 'knowledge utilisation model' and approximates to the 'pull' model previously described. Here success is seen as cumulatively dependent on the prior factors mentioned, but also on complex relationships between users, researchers and group problem-solving approaches. The final model is the communication and feedback model, which is the most elaborate and argues that the requirements already quoted are necessary, but that a feedback loop from users is an additional success criterion. This final model comes closest to a recognition that users' interpretations may vary and affect outcomes. Progressing through the models in this way helps to illustrate the gradual development of understanding of the variables influencing diffusion.

This brief review of current models of diffusion demonstrates an increasing recognition of the complex issues involved. However the underlying concepts still draw on the idea of a 'flow' through the innovation cycle. Albeit in the communication and feedback model, this is not purely a linear flow as there is an acknowledgement of loops in the flow. The inclusion of feedback loops, the data from empirical research on the adoption of clinical guidelines and the evidence on 're-invention' of innovations (Emrick *et al.*, 1977; Rogers, 1995) all raise fundamental questions as to whether diffusion can be understood as either a linear or a 'flow' process. Are innovation adopters really passive receptors of ideas? Research by Jelinek and Schoonhoven (1990), which draws on experience in the private sector, in this case in high technology firms, illustrates a concern with interconnections, overlaps and multiple teams and multiple sets of relationships. One might argue that these ideas and concepts come close to questioning the basic utility and validity of the linear model. Moreover they show greater concern for the interactions between groups of adopters. Some authors have argued that integrative approaches are required to include all aspects of the management of innovation, from developing, to implementing, to institutionalising an idea (Van de Ven, 1986; Burt, 1987; Midgley *et al.*, 1992; Ferlie and Pettigrew, 1996). One concept which emerges from this work is that of networks or communities (and we return to explore this concept further later in the chapter). Networks or communities act as diffusion forums (see chapter 12) with differing stabilising forces holding them together. Such communities may help to diffuse an innovation and to marginalise alternative technologies or concepts.

However we would argue that, not only are linear models poor representations of the process of diffusion, but there are a significant number of other characteristics which are partially understood or underresearched. For example, even in the influential work of Rogers (1983; 1995) a five-stage linear model is utilised and attention is drawn to the role of individual actors. Using the individual in the organisation as the unit of analysis, the characteristics of 'innovators' 'early' and 'late adopters' and 'laggards' are exemplified. Rogers characterises the 'innovation-decision process' as essentially one of choice between an accept on a reject decision. While discussing the concept of re-invention, as a process by which adopters of innovations adapt those innovations, this is represented as almost aberrant behaviour. There is still limited attention given to the impact of the actors' own values and interpretations. They are depicted (still in passive mode) as persuaded to accept or reject the innovation, but they are not seen as interacting throughout the diffusion process, or fundamentally changing the innovation or re-interpreting it.

It is significant that, on the whole, context and the influence of context on diffusion are noticeably absent in most of the analysis. Kimberly (1981) and Kimberly and Pouvourville (1993) are among the few authors who do explore and highlight the influence of context on the diffusion of innovations. This work emphasises the need for contextual analysis and highlights three areas of particular concern and influence in health care: firstly, the degree of centralisation of decision making; secondly, the extent of dissatisfaction with the current health system; and finally the extent to which increasing costs are the driving force for innovations. Broadly this research suggests an additional level of analysis of the organisational context, at sectoral and organisational level, is required if one is to understand the processes of diffusion. There are significant similarities between this research and the research on effective change processes in health care. The role of change champions, the establishment of a supportive climate for change through management and HRM networks of support and the importance of context have all been highlighted as major success factors in research on organisational change processes. Notable in the health care context is the work by Pettigrew *et al.* (1992), because it underlines specific features of the health care context, which are different from those found to facilitate change in private sector contexts. For example, a change leadership group or dyad works best in health care, as compared to a single leader in the private sector. Research then suggests that context is an important dimension of the analysis of the diffusion of innovation and this message is reinforced by the management of change research.

Some of the developing complex models are influenced by other prior research, mainly in the private sector. This draws on a different tradition and is based on different founding perspectives. Here the diffusion of innovation is perceived as predominantly a social process and, moreover, a socially constructed process. Because of the case material used in this chapter, it is interesting to note that, in 1975, Richards drew on the basic concept of scientific work as a social activity to explore the extent to which certain obstetric practices were based on evidence. He concluded that they were not and that they had been influenced by history, assumptions, professional groupings and identification (obstetrics was established as a surgical speciality in the latter part of the nineteenth century!). A focus on the social networks themselves and the identity of the members is a feature of the later work of those authors building on the concept of the actor network (Latour, 1987; Callon *et al.*, 1992) (see also Chapter 14). These authors are seeking to describe and explain the relationship between research and action. Their findings are based on historical and sociological studies of the day-to-day work of scientists and technologists. These studies try to examine the content of scientific work and not just its outcomes. Scientific knowledge is not displayed as neutral and formal, but rather full of judgements and tacit knowledge. Their focus is on diverse interrelationships between networks of heterogeneous actors and materials, who are involved in a continuing process of negotiation to organise and impose their own assessments and meanings. Networks and communities of knowledge are thus seen as key vehicles in the diffusion of ideas and in the gaining of scientific acceptance. Emerging from this work are questions about the passage of an innovation into use. Not only is the linear model of the path of an innovation into use unsupported, but the very thought that every innovation will or can follow a 'model' path into use is questionable. The idea emerges of facts being 'translated' by various actors at various times and in differing ways in order to obtain the buy-in and agreement of others. Translation involves mediation between one community or network and between one context and another. It is perceived to be an important concept in our understanding of the way ideas and innovations diffuse. Ultimately, then, each innovation evolves along a largely idiosyncratic path.

Having explored some of the models and concepts underpinning our ideas on the diffusion of innovations, a number of observations can be made. Almost universally writers and researchers assume that innovation is positive and that therefore people will wish to adopt an innovation. This assumption needs to be questioned and examined. In many instances, the research evidence is perceived as 'scientific' and therefore clear and unequivocal. If this view is accepted then, justifiably, effort is directed towards the clarification and communication of the 'evidence'. With the exception of actor network theory, the diffusion of innovations is portrayed as a complex linear flow. The majority of the diffusion models are context-free and there is an (often implicit) suggestion that one model fits all. In the specific health care context, many of these ideas are employed in the thinking which underpins the development of clinical protocols and guidelines. Applying the models of diffusion in this way leads to a reconstruction of the management and care of patients as a sequence of individual, formal and rational decisions. Furthermore it can create the illusion of a single answer. In this chapter, we seek through our empirical material to question the reality of both of these ideas.

METHODOLOGY

As stated in the introduction, this chapter is based on the findings emerging from two matched studies on the diffusion of innovation in health care. The first study explores the diffusion of four innovations in the acute sector and the second traces the diffusion of four innovations in primary care. The studies were started in 1995 and 1997, respectively, and were of two years, duration, so the second study is still going on. Therefore the material presented in this chapter will largely be drawn from the first study in the acute sector, with some tentative comparisons and reference to the second study in primary care. Each study represents a major empirical project and together they can be seen as a yielding substantial and unique comparisons between the acute and primary health care sectors.

Little prior research has taken place in the UK to explore *why* professionals adopt innovations, which sources and modes of communication of evidence influence them, or indeed what makes evidence credible. As Rogers (1995) states in his criticisms of diffusion research to date, 'We should increase our understanding of the motivations for adopting an innovation. Strangely, such "why" questions about adopting an innovation have only seldom been probed by diffusion researchers' (p.109). Methodologies such as those frequently utilised in the USA, such as the analysis of variance or the medical model of the randomised controlled trial, are not appropriate to answer these questions. In order to explore these 'why' questions we have adopted a qualitative research methodology which allows us to examine the trajectory of innovations and to probe the influences and reasoning of participants.

In each study the definition of 'innovation' was identical and included innovations in technology, clinical methods and the organisation of services. The criteria for selection of the innovations were also identical. In each study, two innovations were included which had strong scientific evidence to support them and two which were supported by contestable evidence. The evidence was adjudged 'strong' on the basis of publications in reputable medical journals, reinforced by peer review by a group of medical specialists who formed our steering groups. The methodology employed was a two-stage one. In the first stage, the diffusion of the innovations across a region was assessed, through interviews with opinion leaders in a range of appropriate medical specialities, including public health and nursing. In the second stage of the study, a micro analysis of each innovation was undertaken by examining the diffusion of the innovation in detail in one specific setting. This stage also included interviews with staff at all levels and in a range of medical, nursing and paramedical professions.

In this chapter it is only possible to provide a brief overview of the innovations studied. In the acute sector project, these were as follows:

- the use of a new computer-supported system to manage anticoagulation service provision;
- the introduction of new service delivery systems for the care of women in childbirth, as specified in the document 'Changing Childbirth';
- the use of laparoscopic surgery for inguinal hernia repair;
- the use of low molecular weight heparin and anti-thrombolytic prophylaxis following elective orthopaedic surgery.

The data presented in the remaining sections draw heavily on the obstetrics case example.

The logic of the methodology is predominately exploratory. However the methodology is designed to produce an analysis of the extent of the impact of strong scientific evidence on the diffusion process. Conceptually the starting point for both studies is to question and critique the current models of diffusion and especially the linearity of many models. Each project has taken a slightly different theoretical stance as its grounding. In the first study, we seek to explore the appropriateness of the concepts of actor network theory to the diffusion of innovation in health care. Specifically the idea of diffusion as a translation process is explored through the empirical material. In the second study, we draw on a framework based on and adapted from the work of Kimberly. Here we are particularly interested in the extent to which the differing context of primary care shapes and remoulds the diffusion processes.

'SCIENTIFIC EVIDENCE' AS A SOCIAL CONSTRUCTION

Our empirical data show that scientific evidence is not clear, accepted and bounded. There is no such fact as 'the evidence', there are simply bodies of evidence, usually competing bodies of evidence. Even the construction of the research agenda and therefore the available bodies of evidence are socially and historically constructed. Obstetrics represents a good example of this.

By comparison with many specialities, obstetrics has a longer history of audit, standardised procedures to investigate perinatal and maternal mortality and established, national databases. More recently it has been one of the first areas for the publication of Cochrane meta analysis (Cochrane, 1996) and has been audited by the Audit Commission (1997). All of this suggests that there is a huge body of available evidence on the effective care of mothers, ante-natally and during labour. This is true, but the available evidence is skewed. The most researched areas are at the extremes of the normal distribution of the population and focus on occasions when pregnancy goes wrong. As one obstetrician put it: 'Women do hop from both high and low risk categories. But there is this grey area. We have black and we have white, but we have 101 shades of grey in the middle.'

The emotional pull of risk is obvious, but the greatest benefit to the population as a whole might be derived from research into the effective (and cost-effective?) care of the 'normal' mother. There are complex reasons for such a research focus never having been adopted. Historically some of these are to do with the growing dominance of obstetrics over midwifery and a professional interest in the difficult and not the normal. The original identification of the obstetrician was with surgery, as obstetrics was defined as a surgical speciality. Thus career progression and the norms of the medical profession skewed research interests. There were also issues of research funding and so on, but none of these could be described as rational or scientific reasons. So the outcome is that there are areas of care for normal mothers, such as the effect of eating during labour, which are unresearched and so there is no evidence upon which to base practice. As a result there are widely differing practices and, indeed, practices and views are reported to have changed over time.

More pertinent to our argument, there are vast areas of contested evidence. There are a number of reasons, some particular to obstetrics, to account for this. The first is the skewed nature of the evidence base, mentioned in the previous paragraph, which means that there are many grey areas, where practitioners have little evidence to guide their practice. Secondly, a commonly expressed concern was that evidence drawn from certain populations of mothers might not be relevant or valid for the population dealt with by this trust or the individual practitioner:

> I do not think you can use all the research papers as evidence.... But I think it is important not to jump on the bandwagon with everything that comes out. And really to review in the context of the women you are dealing with. If you take social deprivation, most of the them [that is the mothers at this unit] are fairly high risk before you begin. You are not dealing with a healthy, well nourished, well housed, high employment population, so you are dealing with a group of women who are socially deprived, large ethnic minority, high unemployment rates, high social problems, overcrowding... so before you begin you have high risk. (A director of midwifery)

This problem exists and is evident in many specialities, but is dominant in obstetrics, because the heterogeneity of the 'patient' population is at its most extreme in maternity care, where virtually any woman can get this 'condition'. So the transferability and applicability of research findings is a major issue.

Another reason which accounts for the contested evidence is the presence of a multiprofessional care group. The history, training and approach to care means that clinicians and midwives do not start from the same foundations and predictably can adopt differing approaches to research evidence. Perhaps most significantly of all there are still only limited data to suggest that cross-fertilisation of ideas between professions is regularly occurring.

One example is selected here as an illustration of a contested area of evidence. The example is the topic of 'active' or managed labour as the safest and most effective approach to labour management and delivery. This body of evidence is contested by the view that, as far as possible, the most effective approach to labour management and delivery is that intervention should occur only as and when required for a medical reason. Predictably, perhaps, one finds that one is not comparing like with like. The proponents of the active labour management school premise many of their arguments on the view that a long labour is highly 'undesirable':

> We work with a rigid straight line partogram and if a patient falls two hours behind the standard, oxytocin is used, automatically. We have now got to the stage where the midwives are empowered to put up the oxytocin. So that winds up in this unit, with a mean length of labour of first-time patients of six and a half hours. I've just examined data from the ———— hospital and it is extraordinary, but there are first-time mothers who are still labouring for 18–20 hours, which I think is barbaric. It comes from the 1940s. (An obstetrician)

The second, more non-interventionist view would tend to start from the premise that childbirth is a natural process in which only a minority of patients will need assistance. Interestingly both schools of thought argue that 'their' mothers prefer their approach. Not only is the evidence contested on this topic, but the practices adopted are widely varied as a result.

While the example quoted here is drawn from the speciality of obstetrics, throughout our research, the data illustrate that there are few (if any) areas of agreed and accepted evidence. The innovation studied in orthopaedic surgery provides a completely different setting, but similar outcomes. We studied the use of low molecular weight heparin and anti-thrombolytic prophylaxis following orthopaedic surgery. These types of drugs prevent patients suffering from clots and heart attacks after surgery. Despite a long history of research over a 20-year period and the involvement of major research centres, the evidence is still very much in debate. There has been meta analysis of the trials, a Consensus statement in 1992, and yet the controversies still rage: 'There are camps. And they are well defined camps in very entrenched positions.' The data suggest that, even after such a long time, indeed partly because of developments over time, there is no widespread acceptance of supposedly 'strong' scientific evidence. To quote one respondent: 'This is a field where an awful lot of RCTs have been done. More than any other branch of orthopaedic surgery. Yet for something which seems to work, why does not everybody use and embrace it wholeheartedly?'

These research data illustrate that evidence is not bounded – it is not clear when the definitive truth has been discovered. So far, we have sought to demonstrate that the construction and interpretation of evidence has to be clearly understood as a social process. It is not a process which can be performed by one isolated individual; diffusing evidence into practice is a construct of debate and of agreement among practitioners.

COMPLEX HIERARCHIES OF 'EVIDENCE'

To add to the complexity, 'evidence' takes a number of forms. There appear to be hierarchies of evidence, where position in the hierarchy relates to the degree of credibility of the source. However the hierarchies are not stable between professions or specialities. Most frequently quoted as the top of the hierarchy of evidence is the randomised controlled trial (RCT), whose results may then be published in a reputable journal. A consultant surgeon said, 'To my mind the proper randomised controlled trials are what counts. If I was writing a paper on hernia repair, those are the things that I would reference.' While RCTs were widely quoted, there was a degree of scepticism. Another respondent, a consultant obstetrician, adopts an unusually critical perspective: 'Because people are throwing trials together. A gospel according to the randomised controlled trial and I am not sure we should accept what the randomised controlled trial says without having a very clear remit.'

Alongside the assessment of RCTs as the 'best' methodology lies an assessment of the credibility of the research results, based upon the reputation of the journal in which they are published. One or two journals, such as the *British Medical Journal* and the *Lancet* (or a few American journals, such the *New England Journal of Medicine*) are seen as having general acceptance in the acute sector, but this acceptance is less evident in the primary sector. Beyond those, most respondents referred to specialist journals in their own field as credible sources of information. That meant that doctors, midwives, nurses or physiotherapists did not, on the whole, share common data sources.

In a number of cases, particularly among consultants, there was more detailed elaboration of a set of criteria which defined 'good' evidence. These demonstrated legitimate concerns about the limitations of research methodologies. Such respondents frequently illustrated a sophisticated understanding of research methodologies.

Published research results which are the outcomes of well-thought-out RCTs are nevertheless contested by other forms of evidence. This evidence is produced through the experience of professional practice. Our data suggest that some specialities give greater weight to their experiential learning than others. From the limited range of specialities in our study, it appeared that surgeons and midwives, both in their different ways, considered the experience of practice alongside research. Among the surgeons, the crux of the argument was that, unlike some disciplines, surgical techniques are skill-based and embodied and not uniformly practised:

> Surgery is not like auditing the results of a drug trial. Surgery is a practical skill, it is like carpentry. It is difficult I think to compare techniques, I do not know how to get round it really. (A consultant surgeon).

Amongst the midwives, a different range of arguments were used. This group was concerned with adopting a holistic approach to the care of the mother and child, and felt that RCTs were inadequate because they rarely took account of all the outcomes, such as the psychological well-being of the mother, which was felt to be particularly important:

> For example, in Cochrane, it says there is no evidence to show that creams help in the problem of cracked nipples. But the women might feel better or experience less pain, so you might want to use one.

> I think it is not just the trials that count as evidence. You have to know whether it was a good piece of research etc... I think the collective wisdom of a group of professionals also counts as evidence. You may not be able to prove what you instinctively know.

Midwives quoted a wide range of examples where the experiences derived from practice, particularly where practice had been reviewed over a period with a large sample, might override research results. These examples included interventions such as pain relief through the use of water, suturing and weighing certain categories of mothers during ante-natal care.

The precedence given to the RCT as the 'best' mode of research has led to some dysfunctional outcomes. The impact of this hierarchy of evidence is to downgrade and denigrate other forms of research. While the status of the RCT may be widely accepted in the acute sector, our newly emerging results in the primary care sector show that GPs express different attitudes and views:

> ...the trouble with most RCTs is that they are imperfect in all sorts of ways. It is encouraging that people try to find out evidence, but it is always imperfect.

> Most people don't want to read through a scientific paper and evaluate it and assess it and make a decision; what they want is a consensus based on what there is, which then provides a simple set of guidelines.

> ...other people draw up the evidence and make the guidelines, the fact that GPs don't end up doing that doesn't necessarily mean that the process has not worked, it may mean the GPs are getting it right and the guidelines are wrong.

There is a widespread belief that the dominance of the RCT as the most accepted and 'respectable' methodology has been a contributing factor in the lack of research in primary care. Many GPs and other medical staff in primary care are sceptical about the applicability of RCTs in primary care; for example, they consider there are major problems with control groups. It is critical for these issues to be addressed if there is to be a growth of EBM in primary care. The notion of 'one size fits all' is difficult to uphold.

To reinforce the argument, therefore, that evidence is a socially constructed reality, this section has illustrated that perceived hierarchies of evidence exist. The criteria for the construction of the hierarchy are not uniform and common to all professionals. There are variations of perception and interpretation which differ by profession and by speciality. Thus the same evidence does not produce the same response in every profession; similarly the credibility of types of evidence will vary among a range of professionals.

DIFFUSION IS CONTEXTUALLY FRAMED

So far in this chapter we have considered the nature of the evidence itself and its contestability and the way perceived hierarchies of evidence influence the interpretation of 'scientific' evidence and professionals' responses to that evidence. In this section we seek to underline the fuzzy nature of evidence and the interpretation of evidence by arguing that the diffusion of evidence and its use are contextually framed. There are many important contextual variables which influence the process and the professional's clinical decision which have nothing to do with the strength or otherwise of the scientific evidence. For example, as a result of GP fundholding and multi-funds, a typical comment was:

> Again if you asked me that question ten years ago, I would have said absolutely not or minimally [that is, GPs being influenced by other professionals], whereas now I think there is much better diffusion of information and a lot more contact between GPs.

Using terms like the 'uptake' of evidence tends to imply that science is the critical influencing factor, rather than one of several factors weighed by the professional in decision making. Our data show that, in many contexts, using research evidence to underpin practice is not the only standard of 'best' or successful practice. Patient choices and social factors are seen as mitigating or intervening variables in these examples.

In obstetrics, a crucial question is what weight should be attached to patient wishes and choices and how these can be weighed against clinical risk. Our data suggest that there is almost a continuum of responses to this question. Such a range of responses leads to wide variations in practice. Some examples are given below:

(a) Listening, providing limited information, but then prescribing to the patient:

> Certainly, if we had a woman who said I don't want my labour to be managed actively, I don't want oxytocin, then we would not give it to her, obviously, but I think a woman's choice is governed by the amount of information she is given and the way that information is given to her... locally women know how things

are; because word spreads amongst them...and I suspect if a woman adamantly said I want to have a normal delivery, no matter how long it takes, then she wouldn't come here.

(b) Supplying evidence, listening, influencing and, if necessary, persuading:

> Unfortunately, the vast majority of women I see are in a high risk group which actually can see no reason at all why they cannot have a very low risk mode of care. I like to think I get through to them. I will go and visit them at home. And explain what their pregnancy is about and why they fall into a high risk category. Why it would not be appropriate to have community midwifery care. And why they really do need to be seen, in this case, by the experts, who are the obstetricians.

(c) Supplying information, influencing, debating and then accepting the mother's decision:

> When I start a consultation with a mother, I actually start by saying to her, have you got an idea how you would like to deliver this baby and she immediately has a view. And therefore I am comfortable with whatever she chooses. If she says, 'Well, I do not have a view', then we have a long consultation over giving her as much evidence as I can about the risks of different approaches.

This consultant gave examples of giving a healthy mother an elective Caesarean section.

GPs also point out that the care they provide to their patients is based on a long-term and sustained relationship. This relationship is usually with both the individual patient and the patient's immediate family. The trust and openness which is part of the interaction play a key role in the GP's ability to function effectively, clinically. GPs weigh aspects of the relationship, patient and family circumstances against the clinical efficacy of a treatment. This might lead to a decision to give 'unnecessary' or potentially risky painkillers or to treat a child with antibiotics which are known to have a limited effect on the condition in question.

> The reason we feel we are prescribing more items is because we have a lot of poor people and we do not refuse to give paracetamol for an illness...we see the pressures of low income here, so we have felt,

socially it would provide a bad response from the practice in this locality if we refused to prescribe for children in situations of poverty. (A GP)

CONCLUSION: AN AGENDA FOR THE FUTURE?

The exploration of our empirical data suggests a need to reconsider the meaning of evidence. Evidence is not perceived as concrete and bounded, but as bodies of contestable and debatable data. Perceptions of evidence vary by speciality, by profession and by individual. So the linear models of a 'flow' of evidence from research into practice are under the first stage of attack. For innovations to diffuse, we need to incorporate individual conceptions of credible evidence and of the underpinning reasons which sustain these conceptions. There is no one accepted view of 'evidence'; individuals and groups strongly defend their versions of reality. Adopters can be seen as active participants, regularly re-interpreting the scientific knowledge. These data contradict rationalistic perspectives and dichotomous accept or reject decision processes about innovations. 'Re-invention' as defined by Rogers is a mainstream activity which might more accurately be described as re-interpretation, since it is not solely customising the innovation. The implications of this and other current work for health care in the future and for EBM in particular are substantial. A selection of the most crucial issues will be raised here.

The diffusion of EBM will not occur without the incorporation of interdisciplinary debate, dialogue and agreement based on consensus. Even the consensus will not last, but will need to be renegotiated regularly. A concern about the quality of the evidence is important, but insufficient. A focus on better, faster, Information Technology (IT) may well produce only minor improvement. Attention has to be given to the forums for debate, to facilitating discussion of the boundaries and the variations in the criteria of credible evidence and to the efficacious processes of translation.

Health care needs to draw on and extend what is already known about the effective means to influence behaviour. To explore and evaluate how the diffusion of evidence can best be achieved will require comparative research across a wider range of professional groups. Alternative interpretations will need analysing and different mechanisms for debate and consensus building will have to be explored. There is a fascinating research agenda in developing our understanding of the 'why' questions; why do professionals count some evidence and discount other? There is an equally absorbing (and extensive?) research agenda in discovering the effective ways of facilitating the translation of evidence between researchers and practitioners and across professional boundaries. These data suggest that a truly multiprofessional approach to health care delivery is still quite a long way off!

References

Audit Commission (1997) 'First Class Delivery', Audit Commission Report.
Berg, M. (1997) 'Problems and Promises of the Protocol', *Social Science and Medicine*, 44(8), 1081–8.
Burt, R.S. (1987) 'Social Contagion and Innovation: Cohesion versus Structural Equivalence', *American Journal of Sociology*, 92, 1287–335.
Callon, M., P. Laredo, V. Rabeharisoa, T. Gonadr and T. Leray (1992) 'The Management and Evaluation of Technological Programs and the Dynamics of Techno-economic Networks: the case of the AFME', *Research Policy*, 21, 215–36.
Cochrane (1996) Database of Systematic Reviews, www.update-software.com/clibhome.clib.htm.
Dawson, S. (1995) 'Never Mind the Solution: what are the Issues? Lessons of industrial technology transfer for quality in health care', *Quality in Health Care*, 4, 197–203.
Deykin, D. and A. Haines (1996) 'Promoting the Use of Research Findings', in M. Peckham and R. Smith (eds), *Scientific Basis of Health Services* (London: BMJ Publishing).
Eddy, D.M. (1982) 'Clinical Policies and the Quality of Clinical Practice', *New England Journal of Medicine*, 307, 343–7.
Eddy, D.M. (1990) 'Practice Policies – what are they?', *Journal of the American Medical Association*, 263, 877–88.
Emrick, J.A. *et al.* (1977) *Evaluation of the National Diffusion Network, Volume 1: Findings and Recommendations* (California: Stanford Research Institute Report).
Ferlie, E. and A.M. Pettigrew (1996) 'Managing through Networks: some issues and implications for the NHS', *British Journal of Management*, 7, March, 81–99.
Field, M.J. and K.N. Lohr (eds) (1990) *Clinical Practice Guidelines: Directions for a New Program* (Washington, DC: National Academy Press).
Grol *et al.* (eds) (1993) *Quality Assurance in General Practice. The State of the Art in Europe* (Utrecht: Nederland Huisartsen Genootschap).
Jelinek, M. and C.B. Schoonhoven (eds) (1990) *The Innovation Marathon. Lessons from High Technology Firms* (Oxford: Basil Blackwell).

Kimberly, J.R. (1981) 'Managerial Innovation', in P. Nystrom and W. Starbuck (eds), *Handbook of Organisational Design* (Oxford: Oxford University Press).

Kimberly, J.R. and G. de Pouvourville (1993) *The Migration of Managerial Innovation* (New York: Jossey Bass).

Latour, B. (1987) *Science in Action* (Cambridge, Mass.: Harvard University Press).

Midgley, D.F., P.D. Morrison and J.H. Roberts (1992) 'The Effect of Network Structure in Industrial Diffusion Processes', *Research Policy*, 21, 533–52.

Mulrow, C. (1996) 'Critical Look at Clinical Guidelines', in M. Peckham and R. Smith (eds), *Scientific Basis of Health Services* (London: BMJ Publishing).

Pettigrew, A.M., E. Ferlie and L. McKee (1992) *Shaping Strategic Change; the case of the NHS* (London: Sage).

Rappolt, S.G. (1997) 'Clinical Guidelines and the Fate of Medical Autonomy in Ontario', *Social Science and Medicine*, 44(7), 977–87.

Richards, M.P.M. (1975) 'Innovation in Medical Practice: Obstetricians and the Induction of Labour in Britain', *Social Science and Medicine*, 9, 595–602.

Rogers, E.M. (1983) *Diffusion of Innovations*, 3rd edn (New York: Free Press).

Rogers, E.M. (1995) *Diffusion of Innovations*, 4th edn (New York: Free Press).

Van de Ven, A.H. (1986) 'Central Problems in the Management of Innovation', *Management Science*, 32(5), 590–607.

Weed, L. (1997) 'New Connections between Medical Knowledge and Patient Care', *British Medical Journal*, 315, 26 July.

Williams, H. and D.V. Gibson (eds) (1990) *Technology Transfer: a communications perspective* (London: Sage).

Williamson, P. (1992) 'From Dissemination to Use: Management and Organisational Barriers to the Application of Health Services Research Findings', *Health Bulletin*, 50(1), 78–86.

14 Value Critical Analysis and Actor Network Theory: two perspectives on collaboration in the name of health

Steve Cropper

INTRODUCTION

Ferlie and Pettigrew (1996) report that, after nearly a decade of policy promoting market-like competition in the UK NHS, there was also a wealth of evidence of co-operative (net)working. The new forms of interorganisational working which they catalogue should not be read necessarily as a fundamental shift in organising principle: indeed, Ferlie and Pettigrew (1996) suggest that the diversification may only represent a temporary, 'unresolved excursion' from the principal organisational form of an integrated hierarchy. A second interpretation suggests that, although there are strong institutional pressures on the selection of organisational form in the NHS, they are unlikely to lead in one direction only: variety is to be expected. A third view is that pure organisational form should not be expected at all. Thus Lowndes and Skelcher (1997) suggest that collaborative partnerships (as one pure form) pass through a natural life cycle in which different modes of governance [hierarchy, market and network] assume a particular importance at different points in time, and in relation to particular partnership tasks. (p.2). A final suggestion, pursued in this chapter, is that, in any organising process, different modes or tendencies can be seen rubbing up against one another, often competing or disrupting one another. This view shifts the argument from an expectation of a single, dominant form to one of sheer variety and tension in organising process.

Whichever version is seen as persuasive, it is undeniable that the fragmentation of organisation form has created a new relational awareness in NHS management and, with it, a more significant sense of complexity and uncertainty about organising processes, as Ferlie and Pettigrew (1996) conclude. Recent policy developments – the strong policy mandate for partnership and wider system thinking and a variety of supporting resources (Department of Health, 1997–1998) – are likely only to reinforce the sense of blurred boundaries and certainties.

IS THE POLICY PUSH FOR COLLABORATION PROBLEMATIC?

While the policy mandate is clear and collaboration, or partnership, is now widely espoused, it is also seen as problematic. There is good reason to be sceptical of the will and ability to collaborate effectively. The consultation paper on public health, *Our Healthier Nation* (Department of Health, 1998), for example, asks, 'What are the obstacles to partnerships at local level and how can national government and local players help to overcome them?' The response could be lengthy (Challis *et al.*, 1988; Wistow and Hardy, 1991; Ferlie and Pettigrew, 1996). Thus Challis *et al.* (1988, p.265) note:

> Like our predecessors, we found inter-agency arenas to be largely characterized by limited and conditional interaction rather than by frequent and free relationships; by attempts to resolve existing problems rather than to anticipate future ones; and by relatively small-scale and isolated examples of 'ad hocery' and opportunism rather than coherent and consistent implementation within some grand design. More specifically, the potential gains of coordination had to compete, as an over-arching agency objective, against such organizational imperatives as budget maximization, maintaining autonomy and professional self-interest.

Two common responses to this diagnosis currently prevail. First, there are calls for good, clear management of partnership arrangements. Clarity of purpose, demonstrable commitment and attachment, robust management arrangements, and clear mechanisms for learning characterise successful partnerships (Wistow and Hardy, 1991). While such desiderata follow from the diagnosis of failings, they represent guidance on the endpoints rather than on the process of management and they suggest that a technical (managerial) fix is sufficient. Challis *et al.* (1988, p.275), faced with a similar argument, resist:

> we are not suggesting that there are any cook-book recipes for successful coordination. What we are suggesting is something both more limited and more ambitious. This is that it is possible to institutionalize a commitment to coordination and, by so doing, gradually create the climate, value-system and incentives which will persuade individual and organizational actors to adapt their behaviour accordingly.

Second, there is a developing focus on the role of social capital (Putnam, 1993) in the management of collaborative working. The argument is that qualities of social organisation such as networks of association, mutual understanding and shared norms, trust and tolerance of obligation may limit and ultimately override the cultural, political and professional barriers to interorganisational collaboration (Ring and Van der Ven, 1994): such qualities should, therefore, be systematically promoted.

Strong networks crossing organisational boundaries clearly are important elements in a reading and practice of collaboration, as Wistow and Hardy (1991, pp.46–7) note:

> jointly managed schemes depended heavily for their inception and success on small numbers of highly committed individuals... enthusiasts with networking and deal-making skills... Unfortunately... they tended to overestimate the extent to which their own commitment and goodwill would be sufficient to guarantee their scheme's continuing success and to underestimate the importance and relevance of basic management disciplines... formal written agreements, accountability and control arrangements... they ignored such disciplines at their peril.

So we have a circular argument: on the one side, arguments for good management; on the other, strong but informal ties. Neither is in itself sufficient either as an explanation of or as a prescription for the development and maintenance of collaborative arrangements. An adjacent tack is taken below which suggests that *any* organising process is likely to be precarious at its initiation, and is likely remain so, if only episodically – there are no single or determinate 'fixes'.

Two theoretical perspectives are drawn together to suggest ways in which collaborative ventures materialise, stabilise and may be sustained, or not. The theories are concerned to depict *purposeful* endeavour. Each stresses the *multiplicity of organisational logics* and seeks to map those logics and their interactions. The two perspectives are characterised briefly below, and then three episodes from a single case study are used to illustrate the form of analysis.

COLLABORATION AND ITS VALUE

The modern sense of collaboration is of a positive, purposive relationship between organisations. Linking for the mutually supportive pursuit of individual and collective benefit, that is, collaborative advantage (Huxham, 1996), each organisation nevertheless retains autonomy, integrity and distinct identity and, thus, the potential to withdraw from the relationship. In a previous paper which sets out the theory more fully, Cropper (1996) argued, following Selznick's (1957) argument about institutionalisation, that it is the judicious management of value attributed to collaborative working which enables the constituent parts of collaborative efforts to hold together. Assuming that actors seek many different kinds of returns, some corresponding ways in which collaborative action might construct and command value were suggested.

While value may be found in systems of simple exchange – services for money, for example – collaborative working is likely to produce value in a different way, through a process of pooling complementary resources. The forms and measure of value attributed to collaborative action are dependent on the manner in which it is construed, negotiated and represented. If collaborative ventures are defined in terms of the types of value they create and represent rather than as mechanisms of simple exchange, their strategies, then, are about collectively and self-consciously building identity, orchestrating purposeful activity and promoting that activity as 'achievement', 'breakthrough' and 'progress'. A 'value critical' analysis thus seeks to offer a method for mapping and explaining the movements, dilemmas and deliberations of the collaborative organising process, its participants and its contexts.

Sources of collaborative value

Comparative studies of modes of organising emphasise such qualities as the productivity, reliability or adaptability, legitimacy and efficiency of different forms of organisation. Such qualities may be grouped together as sources of 'consequential' value – in the sense that, while they are important reference points in evaluation, they are not directly manipulable. Rather they are consequential on other primary or *constitutive* sources of value. While claims may be made about the consequential value of collaboration, they will be founded on more specific evaluations of its constitutive value (Cropper, 1996). This might be judged on the responses to the following questions:

- Is there a purpose to this venture which is meaningful and relevant and to which we (our organisation) can (and should) commit? Do we value the collaboration as an *expression of purpose*?
- Does the collaborative add to our ability to acquire or organise resources to deliver activity against that purpose? Is it or could it be a sound infrastructure in which we should invest? Do we value the collaborative as *an organizing capacity*?
- Does the collaborative fit within the existing framework of organising processes? Is it distinctive? Are its relationships to other processes helpful? Do we value the collaborative as *an element of the institutional framework*?
- Is the collaborative operating collaboratively? Is the process good, fair and open – a principled way of working? Do we value the collaborative for *its conduct*?

COLLABORATION AS AN ACTOR NETWORK

Much organisational analysis is concerned with the processes by which patterns of social interaction arise, settle and change or endure and with the effects of such patterning. Negotiated order theory, for example, is recognised as one useful way of exploring 'the symbolic and perceptual aspects of inter-organizational relationships... the evolution of shared understandings among stakeholders' (Gray and Wood, 1991, p.10).

The analysis of organising efforts which actor network theory (ANT) proposes more resembles one of 'construction and engineering' than of social interaction and construal. Intended to explore how actors pursue their interests from remote locations, it is concerned as much with processes of 'long-distance control' as with the immediate negotiation of order (Law, 1986; 1992). Thus Latour (1983) is concerned to understand 'how a few people gain strength and go inside some places to modify other places and the life of the multitudes' (p.163).

ANT is similar to value critical analysis in suggesting relational and discursive complexity: it is distinctive in insisting on the variety of organisational materials. 'An organization is composed of a wide range of heterogeneous materials. Amongst these, we may count people, devices, texts, 'decisions', organizations and interorganizational relations. These materials are all important' (Law, 1994, p.23). The theory does not set out a substantive typology of materials: it argues that the nature of relevant materials is an empirical matter, locally contingent. It does, however, propose some formal qualities which apply both to the materials themselves, and the networks they form. Thus it is interested in their stability or instability, malleability, durability, fidelity, mobility or distributability, and capacity to represent; that is, to become taken-for-granted, versions of the networks of which they are a part.

The approach draws attention to mechanisms that pull diverse materials together, 'package' them and thus generate more or less stable organisational effects. Thus Law (1992) talks of 'heterogeneous engineers': organising centres in the actor network. ANT is also concerned to explore how and when they fail. There is no presumption of a final or definitive order: instead ANT is concerned to explore modes of ordering. Law (1994) characterises four modes of ordering – 'enterprise', 'administration', 'vision' and 'vocation' – each suggesting a different way of representing and, thus, intervening within organisational patterning and character: 'Administration tells of sustaining due process in the face of the opportunistic irregularity of enterprise... Vision tells of charisma and grace, of single-minded necessity' (Law, 1994, p.79). Collaboration might claim a similar status, as a distinct 'way of seeing' (Cropper, 1996) which competes and interacts with the other modes of ordering. What, then, is its nature?

The response here, which the case study seeks to illustrate, is that the character of collaboration lies in an active construction of mutual interest and interdependence. This is built through, and embodied in, ways of representing a system of shared purpose, associated capacity, a fit within institutional context and a mode of conduct – the four types of constitutive value proposed for a value critical analysis, above. Collaborative ordering draws on and interrelates specific instances and representations of these generalised materials to furnish collaboration.

CASE STUDY: THE PROMOTION OF PHYSICAL ACTIVITY

Recently the scientific case was made for a focus on promoting moderate exercise as a part of everyday living (Health Education Authority, 1994). This gave early legitimacy to what Trist (1983) terms an interorganisational domain, labelled 'promotion of physical activity for health'. This sits alongside, but is rather different from, the promotion of physical activity as a part of sport or fitness. A national task force, established under the aegis of *Health of the Nation*, the national strategy for health improvement in England explored the implications for policy and service delivery (Department of Health, 1996; Health Education Authority, 1995). One conclusion was that local strategies, although they might be initiated by local health authorities, would require support and contributions from many different interests if they were to be adequate and effective. The case study account, in three sections below, tells of a collaborative effort to assess how effort to promote physical activity could be reviewed and developed in one health authority. The three sections reflect chronology, but also relate to particular points of analysis suggested by the theories. The language of ANT points to the processes by which collaborative action is pieced and held together, and within this, value critical analysis relates how the collaborative effort was evaluated and contested.

The first episode recounts how the very idea of a strategy for physical activity promotion was attached to an organisational form, a multi-agency strategy advisory group, and how a complement of other materials was attracted to or resisted it. The second explores the development of mechanisms intended to hold the collaborative work together in the face of competing ways of organising. The third considers the effects of producing and issuing a collaborative strategy.

Episode 1 Setting up for collaborative strategy

The task of creating a local strategy and programme of action to promote physical activity was taken on by the health authority. It fitted well with general responsibilities to pursue *Health of the Nation* target areas and with a local population needs assessment which identified, inter alia, high rates of coronary heart disease. Furthermore, conceived as a multi-agency task, it gave further substance to three developing health alliances between the health authority and its local authorities.

A newly appointed physical activity specialist in the health promotion service immediately sought to convene an advisory group, inviting various organisations to nominate representatives to work with the health authority in preparation of a strategy for the promotion of physical activity across the authority's area.

From the start, the leisure services departments of the three local authorities were seen as essential partners. Leisure services departments provide or contract the most visible points of entry into physical activity and, without their support, and without access to the resources, programmes and users of these services, the idea of a collaborative strategy for physical activity promotion was a non-starter. Other organisations were invited to join the advisory group for similar reasons; for example, a voluntary organisation was already piloting a national scheme to encourage physical activity among older people. Thus a network of symbolic, human and material resources was envisaged.

Many accounts of collaborative processes focus on the process of creating a critical mass of connection. Connecting mechanisms include convenors, mediators or brokers (Gray, 1989) or referent organisations (Trist, 1983). Each represents what ANT would term a 'centre of translation', influencing the ways in which connections between actors, agendas and resources are formed and abandoned. ANT uses the term 'enrolment' to describe the process of connection and of gathering resources 'interposing oneself between the target entity and its preexisting associations with other entities that contribute to its identity' (Singleton and Michael, 1993, p.229). To develop collaboration in this view is to move and shift identification, a process which starts with claims to legitimacy. This can be problematic:

> Although an organization must have some minimal level of public legitimacy in order to mobilize sufficient resources to begin operations, new organizations (and especially new organizational forms) have rather weak claims on public and official support. Nothing legitimates both individual organizations and forms more than longevity. Old organizations tend to generate dense webs of exchange, to affiliate with centres of power, and to acquire an aura of inevitability. (Hannan and Freeman, 1984, p.158).

The first two meetings of the advisory group saw most, but not all, invitees attend. The physical activity initiative sought legitimacy, first, by highlighting the purpose of the venture (more active and, thus, healthy people) secondly, by attaching the initiative firmly to the three local health alliances (as part of the local institutional framework), so strengthening both alliances and physical activity group, and, thirdly, by emphasising the 'hope values' of the package – its potential productive and organising capacity (various developments, including new pathways into and out of local leisure services). A commentary (and intervention) at that time recorded the following observations:

> The two meetings have served to place the idea of a strategy on the table. The first meeting minuted agreements that a strategy should be developed, that it should be action-oriented rather than a 'glossy' and that there was a common interest in community involvement and development as an approach... The second meeting was less well attended, but the Director of Leisure Services of [one of the local authorities] arrived unexpectedly and was central to a useful discussion. [He] ... asked the question about what resources are available to motivate participation in the work... challenged the idea of a joint strategy ... and made clear that any attempt to do more than create a framework strategy would not work – the three councils have different priorities and an attempt to develop a tight strategy for investment would be doomed.
>
> A more concrete project for joint working – a local campaign drawing on and supporting the national Active for Life campaign – seemed to be a collective opportunity and rallying point. The group saw an opportunity for some pooling of efforts and resources – marketing channels, joint bids for small grant funding etc., ...
>
> The Director of Leisure Services was again helpful in proposing a meeting to explore whether there was enough commitment within the Leisure Services departments to make a joint strategy work... [It] requires champions within the local authorities willing to take up the flag of physical activity for health (and quality of life) and to wave it within the authority to draw other interests in: initially, these champions are most likely to be the leisure services directors ... although they are always likely to prioritize politically visible services and services which make money and to work through their leisure facilities. (Extract from a letter from the author to the chief executive, health authority, 21 December 1995)

Episode 2 The search for stability and coherence

Actors will seek to create routes, procedures or forums which draw and hold in materials that will build, stabilise and strengthen the initiative. Enrolment and a strengthening of relationships are served if a geography of obligatory points of passage (Law, 1986) can be asserted. The very idea of moderate physical activity formed such a point of passage: there was frequent reference to it and a process of testing whether or not current services did or did not match up to it. The leisure services departments claimed they were already in the business and meeting needs. The health authority argued that more had to be done. Scientific evidence and national policy defining 'moderate physical activity' were vigorously waved, as were studies showing how and why few people were active. One of the local authorities had tabled a draft sports and recreation strategy which made significant reference to the relationship between active recreation and health and well-being; much was made of this, even in the absence of the authority's representative.

As the group sought to assert this new and rather specific version of physical activity, the representations used by the group were nevertheless inclusive. Over several meetings, members of the group were asked to pool information about local policy, services and resources which would be helpful to the group's planning effort. Data collection forms were created and maps collating the returns were displayed. Star and Griesemer (1989, p.393) explain the tension between assertion of a new structure and inclusion as follows:

> Boundary objects are objects which are both plastic enough to adapt to local needs and the constraints of the several parties employing them, yet robust enough to maintain a common identity across sites... They have different meanings in different social worlds but their structure is common enough to more than one world to make them recognizable, a means of translation... The creation and maintenance of boundary objects is a key process in developing and maintaining coherence across intersecting social worlds.

The very idea of 'physical activity for health' was capable of translation back into any and all of the health agencies, leisure and recreation service providers, transport policy makers, and so forth: it has different significance for each, but may claim and be claimed by each. Such malleability or equivocality is highly supportive of the early stages of collaborative working, where actors are bound only at the margin and the task of collaboration is to connect, align and build on those margins. Boundary objects, then, face many ways and are central to interorganisational collaboration as a mode of ordering. Ideas, data forms, data maps and even meeting venues helped to bind disparate interests together. Some could be represented as forms of real progress – towards purpose, useful capacity and effective collaborative conduct – and so build value and attachment to the group.

But another obligatory point of passage – action orientation – which might have secured early attachment and fidelity to the collaborative group was little developed. Very early in the process, the health promotion specialist bid for monies to run two local schools competitions, to design a poster and an activity for use in schools to promote exercise, but the materials required to produce a coherent package – a champion in education, access to education marketing channels and the money – were not all in place. Collaborative ordering was in competition with administrative rules (the conditions on awards) and standardised, regular procedures (mailshots to schools). In the end, the identity of the group rested on the production of a strategy.

Episode 3   Extending the actor network

The impression from many accounts of collaboration is that organisations either have interests and responsibilities or they are uninterested, they are either 'in' or they are 'out'. In practice, it is certainly more complex than this (Huxham and Vangen, 1998): organisations may enter and leave, or may hover on the edges or in the grey, boundary areas – neither properly in nor properly out. There were a number of the advisory group who were seldom seen. Such ambiguity means that progress can be slow, but also that a wider system can be held together and claimed in ways which are not possible under conditions of clarity and certainty.

Painstakingly all the packages of materials which the group had been creating and collating were finalised and used to fill out the structure which had been proposed for the physical activity promotion strategy. After numerous drafts, circulated to members of the group and followed up for comment, a draft strategy was issued for formal consultation by and on behalf of the multi-agency group to a variety of consultees. The strategy was, in due course, endorsed by the three local health alliance steering groups in turn, and then taken through the formal processes of adoption of the constituent bodies of these alliances: health and local authorities, chambers of commerce, voluntary organisations and so forth. ANT points to the loss or 'deletion' of an actor network as it becomes unified, coherent and stable. In short, the network of materials and interactions disappears from view as its effect becomes unproblematic. However weakly formed, bound and effective the physical activity group had felt, the strategy presented as a coherent representation of the work of a coherent multi-agency group and value was claimed for it and drawn from it as such.

Law (1994, p.140) sketches a conceptual space based on the relative durability of materials. Thus all the thought, talk and action in the group had been relatively ephemeral or inconsequential: 'left to their own devices human actions and words do not spread very far at all' (ibid., p.24). The published strategy and its components provided a means by which professional, role and social networks, resource regimes and available technologies could be enrolled. The endorsed strategy provided a mandate and material resource with which to extend 'connective capacity' (Carley and Christie, 1992), bringing dispersed, peripheral actors into the network. It has been remarkably successful, although not wholly so. Other networks provide and demand competing pathways and modes of ordering which resist the collaborative mode of ordering. Chatting after a meeting at which the local health alliance had agreed formally to support and promote the strategy, the leisure services representative made clear 'This is not the only piece of guidance we are working to'. The strategy has not moved far into that particular network, one which is focused on the contracts placed with leisure service providers, their leisure facilities – the buildings, equipment, the drilled people who run them and provide the services – and their own strategies.

Nevertheless the collaborative arrangement has started to develop an identity and capacity, and from a state of fragmentation now appears a coherent and purposeful actor. There is connectivity, but it should be seen as tenuous. As Singleton and Michael (1993, p.230) observe:

> The roles and identities assigned by one entity to another may suddenly be challenged, undermined or shattered. Where, once, the enrolling actor had organized the obligatory points of passage for others, it finds itself forced to traverse the obligatory points of passage that are dictated by others. And it is not only social others who

intervene: the heterogeneity of the networks means that *any* entity can begin to step out of semiotic character within the network.

CONCLUSIONS

The UK National Health Service is entering a period of intense institutional change, this implicated in a wider set of changes to public institutions. Interorganisational collaboration, partnership and integration are dominant themes – indeed, symbols of the changes. Understanding how collaborative arrangements form, hold together and perform is a critical issue. The research agenda offers a spectrum of choice, from policy and evaluation studies intended to support and guide the development of collaborative working to opportunities to explore theoretical insights into collaborative organising. This chapter is located towards the latter end of the spectrum: the conceptual lens and languages of analysis which the two theories offer draw attention to different questions about collaboration than the wealth of management studies of collaboration.

The contention has been threefold. First, any organising process will consist in a jostling of modes of ordering of which one is collaboration: while it may be dominant, its purity is not assured and it is likely to be precarious. Second, collaborative ordering both shapes and uses representations of shared purpose, associated capacity, fit within institutional context and mode of conduct as its primary materials; these serve to draw and hold things in place, or not. Third, clues to continued coherence, or to breakdown, can be found in the claims of value made in these terms and in terms of consequential value. The tapestry of collaboration, partnership and integration that is emerging as public policy is translated into practice will provide ample opportunity to explore these contentions. The aim should be, not to test their truth value, as such, but to understand the value of the theory as a way of thinking about collaboration and its complexities.

References

Carley, M. and I. Christie (1992) *Managing Sustainable Development* (London: Earthscan).
Challis, D., S. Fuller, M. Henwood, R. Klein, W. Plowden, A. Webb and G. Wistow (1988) *Joint Approaches to Social Policy* (Cambridge: Cambridge University Press).

Cropper, S. (1996) 'Collaborative Working and the Issue of Sustainability', in C. Huxham (ed.), *Creating Collaborative Advantage* (London: Sage).

Department of Health (1996) *Strategy Statement on Physical Activity* (London: Department of Health).

Department of Health (1997) *The New NHS – Modern and Dependable* (London: HMSO).

Department of Health (1998) *Our Healthier Nation: a contract for health* (London: HMSO).

Ferlie, E. and A.M. Pettigrew (1996) 'Managing through Networks: some issues and implications for the NHS', *British Journal of Management*, 7, Special Issue, March, S81–S99.

Gray, Barbara (1989) *Collaborating: finding common ground for multiparty problems* (San Francisco: Jossey Bass).

Gray, B. and D.J. Wood (1991) 'Collaborative Alliances: moving from practice to theory', *Journal of Applied behavioural Science*, 27(1), 3–22.

Hannan, M.T. and J. Freeman (1984) 'Structural inertia and organizational change', *American Sociological Review*, 49, 149–64.

Health Education Authority (1994) *Moving On: international perspectives on promoting physical activity* (London: HEA).

Health Education Authority (1995) *Promoting Physical Activity: guidance to purchasers and providers* (London: HEA).

Huxham, C. (ed.) (1996) *Creating Collaborative Advantage* (London: Sage).

Huxham, C. and S. Vangen (1998) 'Ambiguous, complex and dynamic: an exploration of the contribution of membership structures to the achievement of collaborative advantage in tackling social issues', Strathclyde Graduate Business School Working Paper Series, January, Strathclyde University.

Latour, B. (1983) 'Give me a laboratory and I will raise the world', in K.D. Knorr-Cetina and M. Mulkay (eds), *Science Observed: perspectives on the social study of science* (London: Sage).

Law, J. (1986) 'On the methods of long-distance control: vessels, navigation and the Portuguese route to India', in J. Law (ed.), *Power, Action and Belief: a New Sociology of Knowledge*, Sociological Review Monograph, 32 (London: Routledge).

Law, J. (1992) 'Notes on the Theory of the Actor-Network: ordering, strategy and heterogeneity' *Systems Practice*, 5, 379–91.

Law, J. (1994) *Organizing Modernity* (Oxford: Blackwell).

Lowndes, V. and C. Skelcher (1997) 'Modes of governance and multi-organizational partnerships', paper to the 4th International Conference on Multi-organizational Partnerships and Cooperative Strategy, Balliol College, Oxford, 8–10 July.

Putnam, R.D. (1993) *Making Democracy Work: civic traditions in modern Italy* (Princeton, NJ: Princeton University Press).

Ring, P.S. and A.H. Van der Ven (1994) 'Developmental Processes of Cooperative Interorganizational Relationships', *Academy of Management Review*, 19(1), 90–118.

Selznick, P. (1957) *Leadership in Administration* (New York: Harper & Row).

Singleton, V. and M. Michael (1993) 'Actor-networks and ambivalence: general practitioners in the UK cervical screening programme', *Social Studies of Science*, 23, 227–64.

Star, S.L. and J.R. Griesemer (1989) 'Institutional Ecology, "Translations" and Boundary Objects: amateurs and professionals in Berkeley's museum of vertebrate zoology, 1907–1939', *Social Studies of Science*, 19, 387–420.

Trist, E. (1983) 'Referent Organizations and the Development of Interorganizational Domains', *Human Relations*, 36, 268–84.

Wistow, G. and B. Hardy (1991) 'Joint Management in Community Care', *Journal of Management in Medicine*, 5, 44–48.

15 Organising for Incompatible Priorities

Frank Blackler, Andy Kennedy and Mike Reed

INTRODUCTION AND OVERVIEW

Over the past two decades a series of radical changes has overtaken the British National Health Service. A range of tensions, incoherencies and uncertainties that are not easy to understand or to manage now permeate the institution. This chapter reviews aspects of this situation and, in the context of the new Labour government's wish to encourage responsive public services and collaborative working between NHS-funded organisations, local government and voluntary agencies, suggests how the situation might be conceptualised, researched and changed.

Three case studies are introduced to highlight inherent dilemmas in medical work and the tensions in the NHS as they are experienced by health care professionals. The first conjures up a scene in which 30 or so senior managers attending a team development seminar are bought face-to-face with the inconsistencies and tensions of their working lives. They say their organisation is in a funny state of mind and that this makes the corporate agenda chaotic. They experience the presence of objects (communication blocks, structures) that need to be removed. They feel uneasy. They ask who is in charge, when they are.

In the second case we summon up a hot, still summer's afternoon, where the people of a world-famous cancer speciality group meet. Women nurses sit quietly, their hands folded, as the men doctors talk. The fan makes it difficult for them to hear each other. The air conditioning does not work. The women begin to speak. They talk of young doctors, young patients. A doctor wins affectionate applause as he describes a life-saving treatment for childhood leukaemias that cannot be afforded if other services are not cut. The meeting slowly continues. In the hospital outside it is an afternoon of comings and goings, of living and of dying.

In the third case a doctor, a nurse and a business manager discuss what they call strategic priorities. They speak of the presence of 'the management task'. This is discussed as a series of incompletely connected, yet pressing, duties which resemble each other somewhat, yet differ from each other significantly enough to make each of them unique. The doctor, nurse and business manager talk of tensions in the air between them. There are, they say, 'powerful elements' in the 'present situation'. These elements speak to them of older, deeper struggles, they say, in another narrative of power and domination in health care.

Drawing from a study of the way apparently incompatible tensions are managed in an industrial setting, we feature the importance of detailed studies of events such as these. In many circumstances people are highly creative in the ways in which they deal with paradoxes and difficulties. Studies of these abilities feature the processes that are involved and suggest new approaches to management. We outline the multifaceted processes that enable people to contribute, relate to others, change and improvise regarding multiple priorities, labelling these 'perspective making', 'perspective taking' and 'perspective shaping'. We conclude by highlighting research issues that this analysis suggests.

THE HYBRIDISATION OF THE NHS

Transformational organisational change entails simultaneous, multi-level changes in which existing power bases and relations are radically restructured and core cultural values are substantially reformulated (see Ferlie *et al.*, 1996, pp. 235–7). The difficulties of achieving such changes may make it extremely hard for participants to understand and to manage the events by which they are affected.

Within the NHS, as within many other public sector organisations in the UK, concerted efforts have been made to shift control away from professional bureaucracies towards market-driven mechanisms supported by a new managerialism. This involved an (attempted) imposition of a cluster of changes in ideology, structures and operations (Reed and Anthony, 1993). Key elements included (a) centralised strategic decision making combined with decentralised operational control (Flynn, 1992; Hoggett, 1996), (b) the imposition and monitoring of performance at all levels of service delivery (Smith, 1993; Farnham and Horton, 1993), and

(c) a significant restriction of established professional autonomy consequent upon the implementation of 'neo-Taylorite' forms of work organisation (Pollitt, 1993; Broadbent et al., 1997).

Recent research (Ferlie et al., 1996; Dent, 1996; Clarke and Newman, 1997) has demonstrated that this 'new managerialism' has found expression in various ways. Ferlie et al. suggest that, in practice, no fewer than four 'new public management' models can be identified, in which a common set of core themes (such as tighter financial control, stronger line-management structures, deregulation of the labour market and a reduction in professional self-management) are open to a significant degree of ideological re-interpretation and operational reshaping. They conclude that a simultaneous movement towards tighter vertical line management structures and more decentralised and fragmented modes of network-type service delivery has occurred and that this has produced an ideological and cultural *hybrid* of traditional public service values and business-like orientations.

In a similar vein, Clarke and Newman (1997) have suggested that a picture of 'uneven development, variability and complex articulations of old and new regimes has involved a change in the hierarchy of legitimation of different regimes, through which managerialism and business discourse have become dominant over professional, administrative and political discourses'. The discourse of managerialism has, they maintained, established the overarching ideological and institutional context – the 'master narrative' – within which different forms of the new managerialism have been enacted in different sectors and in different ways. Within the NHS, this managerialist discourse of markets, contracting, consumerism, command and control has been particularly prominent, they maintain, in framing the ideological and political imperatives which must inform organisational change, irrespective of which political party is in power. At the same time, it has encouraged, if not demanded, the shaping and supporting by these general policy priorities of specific moves towards more responsive and localised networks of service provision and assessment. This development, Clarke and Newman point out, has, paradoxically, made the aims of integration and control more difficult to achieve.

The situation is highly complex and somewhat fraught. Indeed, as Ferlie et al. (1996) discuss, one of the crucial areas of cultural change in organisations is the area in which the practice of transformational change has typically been at its weakest. Established cognitive frameworks or 'mind-sets' held by the dominant stakeholder groups of professionals and managers have been resistant to the development of an approach that is consistent with the overall shift towards managerialism in the National Health Service.

Given the size, variety and complexity of the NHS, a degree of hybridisation may, in practice, have been a characteristic of the service in the past. However recent research (Pettigrew *et al.*, 1992; Tilley, 1993; Reed and Anthony, 1993; Dent, 1996) broadly confirms Ferlie *et al.*, assessment that hybrid organisational forms have become the dominant organisational feature of the NHS. A confusing juxtaposition of managerial control, professional power and moral order currently exists within the service in which the resulting organisational forms take on an increasingly 'disorganised' shape and identity. Private sector funding exists alongside public sector accountability; integrated service provision must emerge from within a contracting environment; professional collegiality exists alongside managerial hierarchies; and so on. The variety of structural principles or logics to be found within this unstable 'heterogeneous assemblage' of contracting, commanding, negotiating and so on creates an environment that is antithetic to co-ordinated service delivery.

ORGANISATIONAL TENSIONS AND PERSONAL FRUSTRATIONS

The mix of organisational rhetorics, structures and techniques that is associated with the hybridisation of the NHS would be of little significance if the service functioned effectively. To relate this general analysis to the experiences of work in the NHS, in this section we present short descriptions of three specific episodes drawn from the experiences of one of the authors. Case 1 highlights the feelings of powerlessness that a group of managers expressed when, in the context of the developments reviewed in the last section, they were invited to reflect on their situation. Case 2 features the problems that a multi-disciplinary team had in uniting around achievable actions in the context of resource scarcity. Case 3 reviews the tensions that were characteristic of a newly merged, and dramatically restructured, specialist service.

Case study 1   A senior management development workshop

The events described here took place in a health authority formed about 18 months earlier from two, previously autonomous, organisations. The authority is responsible for purchasing health care services for a population of 250 000 people, with a budget of some £120 m. It now seeks to develop its effectiveness by having a series of 'Senior Management Development Days'. On the first day, a series of 'potential internal change agents' lead or facilitate short sessions, with the primary aim of setting out and clarifying the 'strategic development agenda'.

As the day progresses, it becomes clear that, at least according to their own descriptions, many of these 30 or so senior managers experience their working lives as characterised by incoherencies, inconsistencies and tensions. Conventional distinctions between planning, negotiating, improvising and learning appear to have little relevance to the flexible way in which these managers must approach their tasks. At the same time, however, it appears that the managers wish that this was not so. At the management development meeting they ask for a reduction in uncertainty, to 'be enabled to work in an effective environment' and for 'clearer, stronger leadership'. It is not clear what kind of entity 'leadership' is; the term is used as if it were an abstract quality (released by the activities of one person or of a group such as the CEO or the executive team, perhaps) which will infuse the organisation and permit release from undesirable aspects of organisational life.

The impression of dependency grows. Despite the prima facie convincing reassurances of an obviously caring and dedicated CEO, this group of experienced and apparently capable managers says that it feels unclear about objectives, uncertain about the ways in which decisions are made, and trapped within an incoherent 'rhetorical' world which is 'not real'. Not only is the 'organisational structure' not felt to be connected to the (unclear) key purposes of the organisation, but these senior managers have (or are infected by) 'a state of mind' that renders chaotic the 'corporate agenda'.

The participants themselves explain the sense of uncertainty and incoherence that they feel as they go about their activities in terms of a lack of meaningful community of purpose. Problems they highlight include difficulties associated with rules for decision making and delegation, collaboration across and between directorates and departments, workload management, political influences, the maintenance of confidence, communication, fairness, equity, objectivity and technical robustness.

Although these managers describe their experience as if there were objects (such as 'communication blocks' or 'organisational structures') which stand in need of removal or overhaul, it is clear that statements such as 'We should confront dysfunctional political influences' are ambiguous. They suggest a crisis of autonomy. That the managers believe that there are 'political influences' which are 'dysfunctional' (although it is unclear for whom or for what purposes they are dysfunctional) suggests that the location of authority is not clear. In consequence, the identity of the autonomous subject is also not clear. In answer to the question, 'Who should enact these improvements you mention?' the answer seems to be 'the CEO/the executive team/SOMEONE ELSE'.

Case study 2 Problems in a multidisciplinary management group

In this case we describe a meeting of a 20-strong, multidisciplinary team that is responsible for cancer services and palliative care for a large section of a major city and its rural environs. The group is meeting in the seminar room of a purpose-built facility.

Begun just 18 months before, the cancer services and palliative care that the team provides is acknowledged to be the best possible of its kind in caring for the terminally ill. The team members present in the light, airy room include specialist physicians, nurses and other professionals, along with the business management team. Together they represent the university teaching departments and several interlinked hospitals. The agenda for this three-hour meeting is to review the strategic plan for service provision and service development. The clinical directorate team have employed a facilitator (who is a minister, and head of the chaplaincy services) to ensure that the agenda is worked through in good order.

Just before the meeting begins, the clinical director (who is also a professor in the teaching hospital) remarks privately to a visitor that, 'Although the station is beautiful, it doesn't mean that the trains run.' The clinical director opens the meeting at the invitation of the facilitator and says, 'We have a very thickly packed agenda, and we probably can't get through it in the time. What we are not looking for here are solutions; the aim is to share knowledge. We know that the system is in change, and that it is uncomfortable to be in such a state of flux and uncertainty. Contemporary events (such as political changes and changes in funding mechanisms) mean that the big picture is unstable and as yet unwritten.'

The meeting then proceeds to work its way through the agenda. The first hour is spent on discussing ways of maintaining services at one site while developing new services at another; staffing issues and educational and training issues are to the fore. It turns out that the acoustics of the chosen room are such that it is very difficult to hear what is being said. The nine doctors present have seated themselves in two groups at opposite ends of the long table; during the first hour, at least 90 per cent of the talking is done by the six doctors at one end. They claim that the maintenance of what is nationally acknowledged to be a centre of excellence is dependent on inter-site collaboration. The nursing staff (who run the wards and outpatient services on a daily basis) gradually begin to find their voices during the second hour, and talk about the impact of variability in quality of young doctors in training. They also say that, while they, the nurses, are perfectly able to operate the newly available treatment options, it is the perceived inability of senior medical staff to make quick, responsive judgements in moments of diagnostic ambiguity which causes them most problems, and the service users most distress.

Nurses and doctors agree that, given that the population they serve is predominantly made up of young families, the increase in demand on the service is due to better rates of detection and the availability of new treatments and hence the survivability of several conditions hitherto thought terminal. Noting that there is a particularly high incidence of childhood leukaemias in the ethnic groups served, doctors and nurses agree that demands on their services will grow steadily as the present population ages, even if everything else stays the same.

The discussion turns to money. There is no new money available from the public purse, and awards of money to this service can only be made, it is understood by those present, by the removal of money from other services. The palliative care team say that they are seriously overstretched, and that all palliative care beds are always full. They could use ordinary acute beds, but there is no money to pay for the nurses required, even if they could recruit suitably qualified nurses. The oncology team say that, following a big push forward in health education and screening programmes, their in-patient facilities and their state-of-the-art chemotherapy day centre are at bursting point.

It seems that there is agreement that excellent specialist care is available in institutional settings, but that the demand in the population for treatment of life-threatening illness cannot be met. There is a wish to provide personalised, minimally disruptive treatments in community settings, but pressure on staff means that the service given cannot be as personal or sensitive to individual needs as service users require and health professionals would like to offer.

At this point one of the senior doctors says mildly that the team should start an explicitly aggressive media campaign, publishing the size of the waiting lists and the typical waiting time before beginning treatment by kinds of dangerous cancers. Other doctors join in, saying that they should strike to bring the 'administration' up short. It is clear, however, that no more money will be available under the present government, and it looks unlikely that any government would award significantly more money to this or any other aspect of health care.

A young, internationally recognised specialist in disorders of the blood takes ten minutes to describe a new technique for treatment at the molecular level of certain kinds of leaukaemias. A graduate of a management programme, he costs the development of the new service, and predicts gains in subsequent service use, studiously avoiding any mention of lives saved; he will not be accused of shroud waving. The people present express their appreciation of his work.

On the way into the meeting the author had greeted an elderly, sick man with almost transparent skin as he lay, knees curled up, in the afternoon sunlight. I had smiled and nodded; he had raised his eyebrows in salute. A couple of hours later, an abstract shape was gently lifted into the mortuary van outside the window of the room where the meeting was being held. I could make out his knees, still drawn up.

The meeting slowly continued.

Case study 3   Problems in the organisation of integrated services

In common with many cities that experienced rapid growth in the postwar period, the city which provides the setting for this final case has seen a shift in the bulk of its population from the old city centre to new, extensive suburbs. This has meant that public service facilities – and hospitals are a prime example – are now not close to where the majority of people live. Over the last decade, there has been a slow process of relocating facilities in the new centres of population; this has not been without pain. Ten years ago two hospitals, perhaps called St Xavier's and the Renfrew, were merged, and rebuilt with superb new facilities on a greenfield site; staff still account for things which dissatisfy them in terms of 'those people from the Renfrew' or 'those outdated practices in St Xavier's'.

A new political administration decided two years ago that hospitals should be regrouped in new organisations called 'networks'. This case concerns the network located in the 'growth corridor', so called because it presents the best geographical access for further growth, because transport infrastructure has already developed in response to this fact, and because, for these two reasons, it is the area of the city with the fastest growing population. Five previously autonomous hospital facilities became part of the network; chief officers became general managers, and large numbers of middle managers were redeployed. There was an abiding belief that the management of the network was dominated by the group who ran the largest single component – the teaching hospital mentioned above – and in particular by the CEO of that hospital, who became the network CEO.

In order to enable services to become better co-ordinated, the CEO of the network decided that there should be no more departments, linked as they were to buildings and institutions, but that these should be replaced by programmes. Thus, for example, the Aged Care Programme must be co-ordinated across a specialist aged care hospital, the orthopaedic department in the teaching hospital complex and the psychogeriatric services. Now there are some 13 programmes, and at the managerial core of this arrangement is the clinical partnership. Each clinical partnership consists of a medical director, a director of nursing and a business manager. The link to the CEO and the board is through a determinedly ambiguous role called the executive sponsor. The person in this role acts as an advisor and facilitator, and agrees a service contract with the clinical partnership on behalf of the CEO and board, and monitors performance against agreed targets.

One of these programmes, the Children's Programme, is a complex coalition of different professional groups, led by a young professor, who is determined that his department will rival the world-famous department still situated in the Children's Hospital in the city centre.

The medical director, the director of nursing and the business manager of the Children's Programme held a three-hour meeting to review the strategic priorities of the programme which one of the authors of this chapter attended. As the discussion unfolded, certain characteristics of the nature of these organisational arrangements became clearer:

- In the clinical partners' descriptions of their work and responsibilities the management task presented itself to participants as a series of incompletely connected, yet pressing tasks which resemble each other somewhat, but which differ from each other sufficiently to ensure that they all appear unique – and therefore not amenable to the application of a coherent, accumulated body of 'practice knowledge'. 'You can get better at handling these tasks, at navigating the agenda of problems, but you can't reduce the number or complexity of them.'
- The partners spoke of relationships as being at a given 'stage of development'. They felt that the organisational structures within which they worked and the power relationships between their professions were combining to produce 'artificial rigidities' in their working relationships.
- The participants began to reflect on clinical partnerships in general. These they described in terms of the personal characteristics of particular individuals (and the emergent relationships between them), rather than in terms of historical developments in the practices of medicine or of nursing, and their fiscal and policy environments. Participants acknowledged, however, that the tensions inherent in the clinical partnership arrangements reflect long-standing and deep struggles for dominance in a narrative of power and authority in health care.
- Differing views of the essential nature of health care were evident in the group's talk. Nursing has its roots in a therapeutic relationship of care, while medicine is rooted in a discourse of cure. Two industrial models of the therapeutic process appeared current in the discourse of the group, both of which speak of the location of power and control. A military image of the therapeutic process, as illustrated in nursing, sees control embedded in the structures of organisation (the spatial representation of this model is the hospital, the factory of care, or – more precisely – the ward as the machine shop of care). On the other hand, a professional image of the therapeutic process, as illustrated in medical practice, sees control embodied in the person of the autonomous qualified doctor (the archetypical locus of this image is the operating theatre). Renegotiations at the margins of nursing and medical practice appeared tentatively to be

occurring, although no new term or imagery had yet emerged to help members of the group imagine and develop the possibilities.

THEORISING COLLABORATION IN COMPLEX ACTIVITY NETWORKS

In this section we summarise an approach to organisational analysis that offers a way of understanding processes central to ongoing collaboration in systems that are characterised by tensions and paradoxes. The approach offers a holistic approach to analysing individual agency and collective expertise. As we later indicate, application of the approach helps both to make sense of the episodes just described and to identify processes that enable collaborative activity in complex situations.

Tensions within the NHS reflect a series of unique, and unusually difficult, organisational problems. But all organisations embrace a range of inherent tensions, a point that has been both magnified and exposed by changes to the UK economy in recent years. For example, business organisations that compete in fast-moving global markets must match their short-term needs for survival with the need to resource long-term development, balance established strengths against changing customer demands, undertake advanced planning but at the same time feature improvisation, utilise rules and procedures while fostering dialogue and reflection, and temper the need for individual organisational success with recognition of the advantages of competition through partnerships. Many organisations find that objectives which, in previous times, they would have addressed separately and in sequence must now be met simultaneously. Such situations create many uncertainties and dilemmas and new approaches to organising are needed to help support the heavy demands on people's abilities to collaborate that such pressures create.

A recent study involving one of the authors (Blackler *et al.*, 1999a; 1999b; forthcoming) of an organisation experiencing just such pressures developed one approach to these issues. The organisation studied was a high-technology company whose success depended on innovative and co-ordinated responses by specialist experts who, until recently, had operated in highly segregated ways. In the rapidly changing circumstances of the company, however, staff needed to find ways of reformulating their priorities and reorganising internal and external relationships to make them achievable.

The top management team had no blueprint they could follow to support this reorganisation. Over a two-year period, Blackler *et al.* studied the detail of working life in the organisation featuring the sometimes (although not always) creative ways in which expert groups coped with the competing pressures upon them. By studying everyday practices, the aim of the study was to understand at a general level how people could cope with the tensions, paradoxes and dilemmas of their everyday work, then to feature ways in which such efforts might be better supported. The approach developed borrows from recent developments in the theory of practice, most particularly from activity theory as presented by Engestrom (1987). Activity theory explores how people know and act, introduces the concept of the activity system and features the opportunities for learning and developing that are created by contradictions within and between activity systems. Core features of the approach are as follows:

- Activity theory concentrates on the processes of knowing, doing and collaborating. It analyses these as active accomplishments made possible by the linguistic, conceptual, technological, social and cultural infrastructures which people inherit and help develop. This focus on the processes that mediate and support action stresses the importance of the unique cultural, social, cognitive and behavioural repertoires that characterise different groups.
- Activity theory is distinctive in the unit of analysis it proposes, namely the 'activity system'. A basic representation of an activity system is presented in Figure 15.1. The inner triangle in the figure depicts the central relationships that the approach was first designed to explore: the relationships between individuals, the communities of which they are a part (and with whom they identify) and the activity in which they are jointly engaged and which gives coherency to the whole system. The notion of activity as used here is broader than the notion of 'action' or 'operation' yet more restricted than the notion of 'culture', being close to the notions of 'social script' and 'frame of reference'. The outer triangles in the figure represent the relationships that mediate individuals, activities and colleagues. As indicated in Figure 15.1, concepts and technologies mediate, that is they do not passively support but they actively enable, the relationships between the individual and his or her activity, social rules mediate the relationships between the individual and the work community,

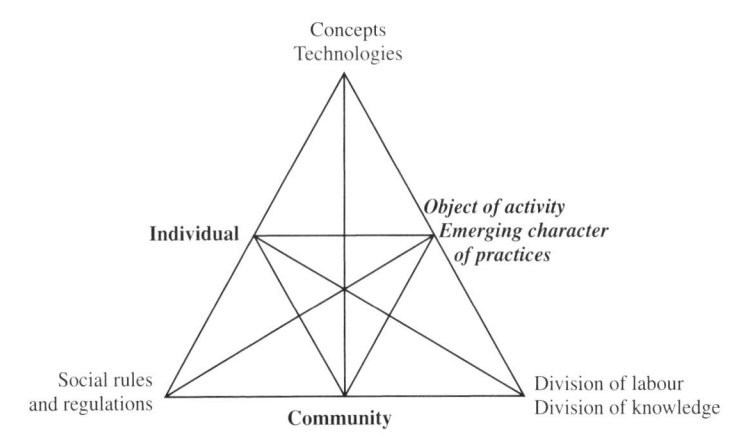

Figure 15.1 An activity system or 'community of activity'
Source: Adapted from Engestrom (1987).

and a division of labour mediates the relationships between community members and their shared activity. In this way the figure represents the technological, social and structural infrastructure of expertise.
- Applied to organisational analysis this approach highlights how people achieve their knowing and who they achieve it with. It provides a holistic approach to analysing the paradigms of practice that distinguish different professional and technical specialists, functional groups, multidisciplinary project groups, management committees, and so on.
- Within a formal organisation, self-supporting 'communities of activity' are emergent, overlapping and interacting. Organisations can thus be analysed as distributed and decentred systems of knowing: that is, as networks of communities of activity (or, for simplicity of expression, as activity networks).
- The infrastructure and priorities that make it possible for particular communities to focus on shared goals, to develop an identity and to act competently can also act as barriers to close collaboration across different communities. Interactions across activity networks are easier to manage in times of relative stability when systems can be developed to minimise conflict or misunderstandings, and more difficult to achieve in times of uncertainty when priorities, methods and group identities may need to be re-examined. In these circumstances

(as in the NHS) shifting relationships between different communities of activity can raise problems about the overall coherence of their activity, appropriate methods, the identity of communities and their relative authority and influence.
- While Figure 15.1 maps the collective infrastructure of expertise within particular activity systems, Figure 15.2 models the core processes of organising within an activity network. As illustrated in the figure, the sense-making processes through which conflicting pressures and priorities are resolved in complex activity networks can be characterised as 'perspective making', 'perspective taking' and 'perspective shaping'.

   (a) *Perspective making* refers to the various contributions that different communities of activity bring to the organisation. It involves the management of priorities, identity and infrastructure. The processes of perspective making support 'domain innovations' within specialist expert communities.
   (b) *Perspective taking* refers to relations between communities of activity and involves the management of influence, both horizontally and vertically within an organisation. Perspective-taking processes support 'boundary innovations' between communities.

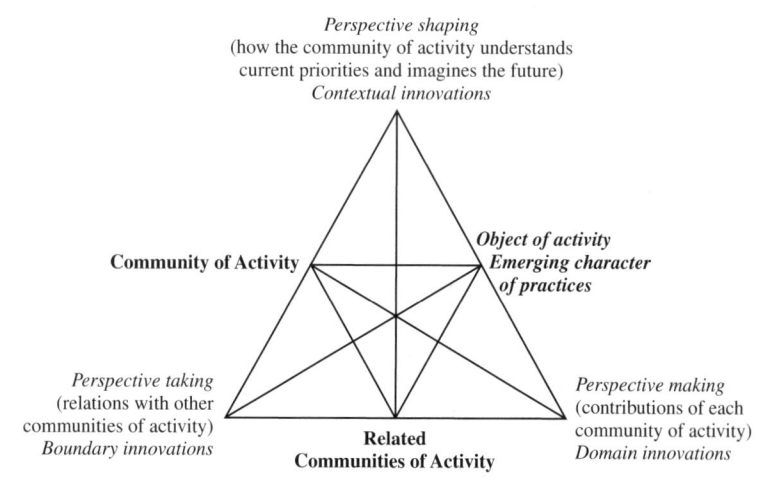

Figure 15.2 Organising processes in activity networks
Source: Blackler *et al.* (1999a).

(c) Finally, *perspective shaping* refers to assumptions about context, achievements and possibilities. It involves the management of the imagination that underlies frameworks of problem identifi-cation, the conceptual resources they bring to their activities and the facilities they have for reflection. Perspective-shaping processes support 'contextual innovations'.

- Typically activity systems and activity networks are, in many organisations, becoming more complex, interrelated and abstract. 'Expansive learning' within activity networks occurs when, in response to tensions and incoherencies within or between such systems, members of a community of activity develop new approaches to their work, define their identities and relationships to other communities in new ways and, collectively, develop new understandings of the context in which they are collaborating. (This is not unlike the process of controlled change and incremental learning that Daft and Weick (1984) dubbed 'enactment'). Effective collaboration in complex activity networks involves a continuing cycle of such questioning, search, development and consolidation. The anxieties and defences that this can provoke may frustrate such learning.

DISCUSSION OF THE CASES

The three cases presented earlier raise a range of interrelated issues that are often dealt with separately. The activity theory approach supports a more integrated approach that features the ways in which activity systems are changing in many organisations at the present time and which focuses attention on the processes that support or inhibit collective learning and creative responses.

One key feature of all the cases that activity theory underlines is the commitment of all those involved to their activity. However case 1 features the way people are likely to feel disempowered, pessimistic and confused in the face of confused priorities and an ineffective and unintelligible activity system. Perspective making was inhibited as members of the group defended themselves by blaming others. It is as if they felt that their discretion for action had been compromised by the need to work in concert, even though they acknowledged that their official mandate required them to work with other agencies and actors towards improvements in health care for the population they served.

The demands on the group to accommodate uncertainties and inconsistencies in the work of their organisation have, it would seem, outstripped their capacities to adjust. Although the managers intuitively realised that enhanced intimacy and communication were required ('we want to be open and frank'; 'we don't want to be judged'; 'we want to avoid damaging personal commentary') they believed that a good number of conditions must be satisfied by *others* before they themselves could be expected to act in the new mode.

Case 2 features some of the difficulties of building an activity system that supports cross-disciplinary working; that is, that required the development of an activity system to bridge boundaries across professional groups that utilise contrasting activity systems in order to support effective perspective-taking processes between them. The sense of identity emerging in the group described in this case reflected the dilemmas of its activity and was characterised by its sense of struggle. Much of the tension between the doctors and the other professionals experienced at the beginning of the meeting can be understood as relating to evolutions in the division of labour and knowledge in the team. Certain members experienced proposed changes as encroachments on their professional domains (and hence personal freedoms), or as harbingers of dangerous downstream changes in quality. Nonetheless, in this episode, shortcomings in the understandings and sensitivities of the various participants could be reinterpreted by them as matters of experience, and so remedied over time as participants learned more about the situation and what each of them could bring to it. In fact the team appeared to achieve some success in recontextualising its objectives and in beginning to develop a collaborative approach. Decisions about resource allocation in absolute terms and in terms of the distribution of available resources were understood better in the group when the specifics of their problems were reinterpreted within the broader context of wide-scale social changes that were affecting the activities of them all. Individuals such as the haematologist described in this case find themselves in a tense relationship between what is technologically possible and what is socially regulated. As perspective-shaping processes were supported by informed debate within about specific possibilities, tensions from the overlapping roles and contributions of the members of the emergent community of activity became more tolerable and discussible.

Case 3 describes a situation where old demarcations had been demolished in an ambitious series of structural changes. A new division of expertise and control, built around sophisticated working relationships across previously distinct professional groupings, was needed if the rhetoric of delivering 'the best possible service to the people of this part of the city' was to be met by reality. In truth, however, the range of structural changes that had been imposed had failed to support the emergence of a new, shared activity system across different professionals. Clinical partnerships need, of course, to mediate between the concept of corporate responsibility and the reality of the 'front-line' issues of care giving, rostering, education, quality assurance and so on. The tensions inherent in the work of the clinical partnership described in the case were profound, including as they did tensions between demand and resource, technological capacity and meaningfulness, life and quality of life. In such circumstances the search for simplistic solutions is a recipe for problems. For the tensions in the partnership (temporarily) to be resolved, participants needed, through a process of expansive learning, to find ways of tolerating the anxieties, ambiguities and animosities that the demands for a range of domain, boundary and contextual innovations were precipitating.

IMPLICATION FOR ORGANISATIONAL BEHAVIOUR RESEARCH

Following the election of the new Labour administration in 1997, there was widespread euphoria among professionals working in the NHS. Now (a year later at the time of writing) however, many acknowledge that easy alternatives to resource constraints, restrictions on expenditure and so on are simply not going to be found. Many of those charged with the management of the service find that they feel doubly aggrieved that the new administration slights them in public (and insists on further reductions in management costs) while at the same time it relies on them to deliver a new system of collaboration for which there is no model anywhere in the world. Primary care groups PCGs might turn out to be an exemplary means of facilitating citizen-focused collaboration, but each of the triumphs PCGs achieve involves a struggle with entrenched and restricted views. Managers need support in fostering expansive learning, yet they are consistently required either to make (or to encourage others to make) domain, boundary and contextual innovations, while at the same time they have to respond to central initiatives that have the (no doubt unintended) effect of re-emphasising *existing* arrangements. (For example, certain sorts of initiatives to reduce hospital waiting lists, at time of writing a government priority, cut across other innovative developments, privileging the work of, say, orthopaedic surgeons over the boundary-transgressing work of primary care mental health teams).

There is no prospect of easy solutions to the many tensions and dilemmas associated with health care provision in the UK at the present time. Struggles for power and control will not simply fade away, for example, because it is now widely acknowledged that a more co-ordinated, less 'market-driven' approach to service delivery is required. Nor will integrated relationships within and between agencies emerge because politicians decree that they should.

Yet, while expanded priorities, revised identities, new relationships with others and new working methods cannot be commanded, their emergence can be supported. This insight promises a basis for new understandings of health care provision and new approaches to its organisation. Despite the many difficulties staff within the NHS experience in their work, it should not be overlooked that staff often (and routinely) cope exceptionally well, demonstrating sophisticated understandings of the complexity of their situations and powerful responses to paradox and dilemma. These abilities need be understood better than they currently are and actively developed and supported.

The conceptual approach outlined in this chapter offers, we think, a promising framework for approaching this task. Activity theory helps explore complex co-operation, it features the ways in which systems of knowing and doing are changing, it emphasises that activity systems are *tension-producing* systems and it suggests that the everyday tensions within and between them are the driving force for collective learning. The framework that we have introduced in this chapter needs further development and application, in particular with regard to the following:

- Further development is needed of the activity theory approach itself, most importantly in an exploration of the nature of 'expansive learning' and a theorisation of the conditions under which it is likely to occur.
- The general framework should be applied in a detailed analysis of actual health care practices, moving beyond the introductory level

of the present discussion. Ethnographic studies are needed to illuminate perspective-making, -taking and -shaping processes and their interrelationships.
- Research is needed to explore the practical utility of the approach in supporting health care professionals in their work. One advantage of the formulation introduced in this chapter is the way in which it frames the complexity of working relations but does this in a way that renders them discussible and, therefore, more manageable. Action research studies are needed to explore ways in which the framework can be used to help people understand and better address difficult organisational problems such as can arise, for example, in the organisation of multidisciplinary groups of professionals, inter-agency relationships, and the relations between central government and local management teams.
- Although they are much used, the terms 'management' and 'collaboration' are poorly understood in the NHS. Activity theory offers considerable potential for the development of an approach to these matters that can be fashioned from an understanding of the realities of the situation and a recognition of core organising processes. What is required, we suggest, is nothing less than the development of activity systems which support the simultaneous achievement of incompatible priorities.

References

Blackler, F., N. Crump and S. McDonald (1999a) '*Organising Processes in Complex Activity Networks*, Department of Behaviour in Organisations Research Paper, Lancaster University.
Blackler, F., N. Crump and S. McDonald (1999b) 'Organisational learning and organisational forgetting: lessons from a high technology company', in M. Easterby-Smith (ed.), *Organisational Learning and the Learning Organisation* (London: Sage).
Blackler, F., N. Crump and S. McDonald (1999c) 'Managing experts and competing through innovation: an activity theoretical analysis', *Organisation*.
Broadbent, J., M. Dietrich and J. Roberts (eds) (1997) *The End of the Professions? The Restructuring of Professional Work* (London: Routledge).
Clarke, J. and J. Newman (1997) *The Managerial State* (London: Sage).
Daft, R. and K. Weick (1984) 'Toward a model of organisations as interpretation systems', *Academy of Management Review*, 9, 284–95.
Dent, M. (1996). *Professions, Information Technology and Management in Hospitals* (Aldershot: Avebury).

Engestrom, Y. (1987) *Learning by Expanding: An Activity Theoretical Approach to Developmental Research* (Helsinki: Orienta Consultit).

Farnham, D. and S. Horton (1993) *Managing the New Public Services* (London: Macmillan).

Ferlie, E., L. Ashburner, L. Fitzgerald, and A. Pettigrew (1996). *The New Public Management in Action* (London: Oxford University Press).

Flynn, N. (1992) *Structures of Control in Health Management* (London: Routledge).

Hoggett, P. (1996) 'New modes of control in the Public Service', *Public Administration*, 74, 9–32.

Pettigrew, A., E. Ferlie and L. McKee (1992) *Shaping Strategic Change* (London: Sage).

Pollitt, C. (1993) *Managerialism and the Public Services*, 2nd edn (London: Macmillan).

Reed, M. and P. Anthony (1993) 'Between an ideological rock and an organisational hard place', in T. Clarke and C. Pitelis (eds), *The Political Economy of Privatisation* (London: Routledge).

Smith, P. (1993). 'Outcome-related performance indicators and organisational control in the public sector', *British Journal of Management*, 4(3), 135–52.

Tilley, I. (ed.) (1993) *Managing the Internal Market* (London: Paul Chapman).

16 Evaluating Interventions to Health Organisation

John Øvretveit

INTRODUCTION

The subject of this chapter is research into health management and policy interventions. It considers why there is relatively little research into organisational changes, management methods and health reforms. It notes the research problems and designs which have been used to evaluate quality methods and which can be used to study other interventions. Finally the chapter considers how managers can gain access to and assess research which could help them in their decisions. The chapter does not argue for evidence-based management, but does propose that managers and policy makers make a greater use of evaluation research. It proposes that the reasons they do not are as much the fault of researchers as they are due to the politicisation of health and the nature of decision making in health care.

The lack of pre- and post-evaluation of many health reforms has often been noted in the health service policy literature. Less remarked upon is the lack of evaluation research into different management methods and organisational changes. Grol notes, 'Research on many interesting strategies is lacking. The effects are largely unknown of organisational development, team building, re-engineering complex care processes with many care providers involved, enhancing leadership in institutions, changing tasks and responsibilities of care providers, or introducing specific financial incentives and economic policies on causing changes in practice performance' (Grol, 1996, pp.239–40). It is sometimes assumed that, if these methods work in the commercial sector, they will be effective in and transferable to health care, even though the evidence of their effectiveness elsewhere is sparse (Pollitt, 1996).

Even more striking is the lack of evidence about and research into quality methods and quality programmes. This is surprising for two reasons. First is the amount of time and money spent on quality activities: what was the basis for this commitment of resources, and has the money been well spent? Second, the quality movement emphasises a fact-based approach and that evaluation is one quality method and an integral part of a programme. If evaluation and self-review are an essential element of quality, what have the research and self-evaluation found? Has the quality movement developed methods for assessing the effectiveness of quality methods which could be used to assess the effectiveness of other management technologies?

As regards audit in the UK, it was reported that:

> The Department of Health is still unable to assess the benefits of clinical audit five years after it was first set up in the health service, the NHS chief executive admitted last week... Some MPs expressed astonishment that the NHS executive has still not measured the outcome of the estimated 100,000 clinical audits carried out by Trusts, health authorities and GPs. A labour MP demanded to know how the NHS could justify spending £279m to date on clinical audit in hospitals – equivalent to recruiting 1,500 doctors a year. (*Health Services Journal*, 21 March 1996, p.7)

Is the situation different in the USA, where many hospitals are a few years ahead of the UK in their quality programmes? Are the promised fruits of these programmes now evident? A study of total quality management (TQM) in US hospitals commented, 'Although there is a growing descriptive and prescriptive literature... no systematic evidence exists as yet to demonstrate CQI/TQM's superiority to existing or alternative approaches to quality assurance and improvement' (Shortell *et al.*, 1995). This conclusion is similar to one from a review of 126 reports of TQM programmes: 'despite a considerable body of literature, very little data exist confirming the claims made on behalf of TQM, including improved performance, quality or competitiveness. No comparative research has been published, and with few exceptions the numerous case studies are anecdotal' (Motwani *et al.*, 1996).

Is the lack of evidence due to poor research? Is it due to problems in evaluating interventions like quality programmes or management technologies? If more and better evaluations were done, would managers and policy makers make more use of the research evidence?

One hypothesis of the present chapter is that research into quality methods would be one area where we would expect to find evaluation methods which could be used to evaluate other management methods and change interventions. Another is that we might learn how to enable managers and policy makers to make a greater use of research findings if we look at the way managers who have been converted to the quality approach have made use of evaluation research into health service quality.

The chapter first considers the problems involved in evaluating hospital quality programmes. It notes some of the research designs which have been used, and the increasing amount of accessible research into the effectiveness of quality programmes and other types of management interventions. The chapter then describes how managers and their advisors can assess and use the research, and notes the implications for future research.

EVALUATING HOSPITAL QUALITY PROGRAMMES: PROBLEMS AND DESIGNS

For clinicians, and for many others in the health sector, experimental principles and the randomised controlled trial are the model for an evaluation, and evidence from other types of evaluation is viewed with suspicion. This approach considers the treatment to be evaluated as an experiment. Patients are randomly allocated either to the treatment or to a placebo and patients in both groups are measured before and after. Statistical analyses are performed to discover whether any difference in outcomes between the two groups is greater than that expected by chance. These analyses and careful controls aim to rule out any explanation for the difference in outcome being due to factors other than the treatment, and to establish whether there is a causal influence at work.

Health personnel, who are one set of 'customers' of a quality programme, expect evidence from evaluations of quality programmes. When they ask for evidence many have in mind the evaluation design of an experimental controlled trial. They find research which does not use experimental design to be less credible, and more difficult to assess if they have not been trained in social science or policy research. Problems using experimental designs to evaluate social interventions have been much discussed within the multidiscipline of evaluation (for example, Shadish *et al.*, 1991; Smith and Cantley, 1985; Øvretveit, 1998). A hospital quality programme poses even more challenges to the experimental evaluator than do many other social interventions. The intervention is difficult to specify, and the details of each hospital quality programme are different. The programme is usually different from the planned intervention and is often changed as it proceeds. Many evaluation reports can be criticised for not giving a good description of the hospital quality programme, but this is not just a failing of the report: skilled researchers who wish to describe quality programmes face a number of problems in identifying exactly what the intervention is or was.

The state of the organisation before the intervention is also thought to be an important factor in the likely 'success' of the intervention. Treatment evaluations involve controls which screen out patients with co-morbidity and history which confound the evaluation. In contrast, for quality programmes and other management interventions, there is little understanding of the pre-existing factors and conditions which would predict success and failure, and less ability to control for these. Consequently it is difficult to select comparable 'cases': the fact that two or more hospitals have a similar number of beds and employees says nothing about their capability and readiness to take on a quality programme.

Further problems in applying full 'treatment' experimental evaluation designs are that the circumstances surrounding and influencing the intervention often change significantly during the evaluation and cannot be controlled to rule out confounders. A hospital may start a quality programme and then a health reform may be introduced, as with the evaluation of the NHS quality programme (Joss and Kogan, 1995), or there may be radical changes in the financial or competitive climate, or the management may change (Øvretveit, 1999). Even if a similar hospital which does not have a quality programme is used as a case control, it is difficult to ensure that both hospitals experience the same internal and external changes which might affect the measures of outcome which are used in the evaluation.

There are also problems in defining what would be signs of a successful quality programme or acceptable outcome measures. Although higher patient satisfaction and better medical outcomes are generally viewed as the aims of quality programmes, they are often not the only aims. In recent years many in the USA have come to view cost reduction as an important, if not primary, aim. US hospitals seek increased market share and profitability, while some other hospitals, such as Norwegian teaching hospitals, put more stress on improving their image so as to recruit scarce staff or win research contracts or finance for new developments. Not only are there different aims but there are also different views about how to measure outcome. Furthermore many quality experts propose that patient and financial outcomes will only be detected three to five years after the programme has been started. In principle this means a long-term evaluation which documents changes which may affect the measured outcome, apart from the quality programme. The longer after the intervention that the outcome is measured, the more likely it is that factors other than the programme cannot be ruled out as explanations for the outcomes.

These features of the intervention (the quality programme), of the object of the intervention (hospitals) and of the environment surrounding the object (the political, financial and other circumstances affecting a hospital over a period of years) cannot be controlled and make it difficult to be sure that measured outcomes are solely produced by the quality programme. These problems make it difficult for others to be sure that they would get the same results if they used the same intervention, even if the evaluation gives a clear description of the details of the particular programmes which were evaluated. (See Box 16.1.)

Box 16.1 **Challenges for researchers studying management technologies and interventions**

- Many management interventions are poorly specified, and/or multiple-component.
- The mechanism of effects may be a systemic interaction between components, but the nature of this mutually reinforcing interaction is not well understood, and there is little predictive theory for the intervention which would be necessary for an experimentalist approach.
- The intervention and context changes.
- The intervention is often unique and the objectives of the intervention are often different for different organisations.
- The pre-intervention state of the organisation is often important to the success of the intervention, but the critical success factors may not be well understood for health care organisations.
- The organisation changes and so do comparison organisations (for example, 'controls').
- It is difficult to measure effects and some effects, especially unintended ones, may only become apparent after many years.

These problems of control and lack of theory do not mean that experimental principles should be abandoned entirely in favour of descriptive 'process' designs, with, at most, speculations about causality. Similar challenges face those evaluating health promotion programmes, and evaluations in this field have developed methods, concepts and models of mechanisms which can be drawn on to improve evaluations of quality programmes and other management technologies (Macdonald et al., 1996). The problems do call for more attention to developing methods for evaluating management technologies and to methods for communicating the results to managers and policy makers.

Designs used to evaluate hospital quality programmes

The difficulties using experimental and even quasi-experimental designs have not prevented researchers and quality specialists from trying to evaluate quality programmes. Different approaches have emerged within evaluation in response to the criticisms of experimental approach: the 'neo-experimentalists', some of whom use qualitative methods, the phenomenologists and political scientists who examine social process and explore the meanings which different interest groups give to interventions, and the action researchers who aim to help those applying the intervention (Øvretveit, 1998a). These and other approaches lie behind the designs which have been used to evaluate quality programmes.

A review of published evaluations of hospital quality programmes, carried out as part of a three-year evaluation of quality programmes in Norwegian hospitals, found that most designs could be classified as one of ten types: surveys, single case descriptions, single case audit; multiple case descriptions, multiple case descriptions with case control(s), single case, with evidence of effects, multiple case, with comparison of effects, multiple case with control(s) and comparison of effects, multiple case, with comparison of standard measures of effects, and action evaluations. Methods for internal evaluations by managers are described in Øvretveit (1998a). One useful design was an action evaluation design which involved the researchers assisting managers to collect evidence about their quality programmes in a standard form for their own use. The evaluation then checked and used the evidence which each hospital collected for their self-evaluation as part of the overall comparison of the hospital's programmes (Øvretveit and Aslaksen, 1999).

> **Box 16.2 Questions to ask to make a quick assessment of an evaluation of a quality programme or project**
>
> - Relevance and replication: does the report give a description of the quality programme and health care organisation which is sufficiently comprehensive and detailed to allow others to judge the applicability to their own organisation and to repeat the intervention?
> - Evidence of outcomes: does the report present data about changes to patients or providers which might be attributable to the programme?
> - Certainty of attribution: does the description, the design and the conduct of the evaluation allow us to infer reliably which factors caused the outcome?
> - Self-criticism: did the evaluation discuss methodological issues, specify assumptions, list the limitations and define the scope of applicability of the conclusions?

TOWARDS EVALUATION-INFORMED MANAGEMENT

If more research into organisational changes and management methods is carried out, how might managers and policy makers make use of this research? Management is just as pressurised as clinical work, so any action to increase the use of research in management decision making has to be practical and to recognise the conditions under which management decisions are made. This section of the chapter considers the four steps which managers can take to make a greater use of evaluation research. An understanding of how managers might use research is important for researchers and for research funding bodies to decide which research to undertake and how best to make it accessible to managers. The four steps described are defining the question, searching for research, assessing the research and deciding the local implications. Managers can delegate some of this work, but they need the knowledge and skills to be able to do it themselves.

Spending time deciding the right question and making it specific and answerable is a necessary first step to guide a search for the relevant research (Øvretveit, 1998b) as Stewart also suggests in her Foreword to this book. It also focuses the manager's attention on what exactly is the issue: being able to take a general inquiry or area of interest and to define an answerable question is an important general skill in management. 'How do we improve co-operation between primary care and the hospital?' is a general 'marker question': it marks out an area of interest, within which work then needs to be done to decide the specific questions which will guide the research search.

To specify the question we can ask: which are the most important aspects to improve (for example, cooperation), which aspects may have been studied elsewhere and which aspects might be improved by knowing how successful different actions have been when carried out elsewhere? A question such as 'what is the most cost-effective way to ensure that general practitioners and nurses are warned about a forthcoming discharge and get details as soon as possible?' can be answered meaningfully.

The search for relevant research studies is probably the most time-consuming step, and one which managers might delegate to assistants or specialists. Unlike the databases of research for clinical disciplines, management databases have been poor, difficult or costly to access and there have been few good electronic databases (the Leeds University 'HELMIS' and the NHS Centre for Reviews and Dissemination databases are exceptions to this). In addition, the design of a study is often not well described in abstracts. There are fewer reviews of research and the searcher usually has to go direct to the primary research study, and often needs a copy of the full research paper to decide whether it is relevant or not. Time can be saved by asking experts for their advice about whether there are studies relevant to the question, and by following up references listed in the most recent research.

The practical problems of discovering relevant research are equalled by the technical problems of assessing the quality of the research and judging the certainty of the findings. This raises the questions of what constitutes 'evidence' in management research and whether ordinary managers can assess the validity of management research. To do so properly requires an understanding, not just of experimental evaluation techniques, but also of social and policy research methods and of qualitative data-gathering methods and analysis. For some types of research, managers will need expert researcher advice. However the author's experience training managers and others is that they can make a reasonable assessment of most types of study if they use frameworks to guide their assessment. The first is a general purpose assessment framework:

Assessing an evaluation: score 1–5 for each
1. Description
   Is the intervention well described? (Described over time – any changes in the intervention described; the elements of the intervention? Intervention separated from other change/things? Replicable?)
2. Data
   Are the measures or other types of data true or an artefact of the researcher or method? (Valid? Reliable? When? How long for? How often?)
3. Design
   Could anything else explain the findings? (Are all confounders considered? Which are controlled for? Does the design exclude other explanations? (Were these predicted at the start of the study?) Does the analysis consider and account for other explanations?)

The second approach is to grade the design used in the research study for its potential to produce valid findings about the outcomes of the intervention in question. Those designs which are graded lower in the hierarchy can be excluded and conclusions from those graded as producing acceptable evidence can be drawn by looking for any patterns in the findings. The type of grading system to use depends on the type of intervention which is the subject of the research. A single 'hierarchy of evidence' which applies to all management and policy research is not appropriate. The hierarchies used in evidence-based medicine are based on experimentalist criteria where outcomes are studied in a particular way and the assessment is made in terms of the way confounding influences are controlled for (Fowkes and Fulton, 1991).

For some types of management research, experimentalist hierarchies can be used. However, in much management research, controls of the same type are not practical or ethical, and an experimentalist approach with a narrow focus on one or a few measurable outcomes is not appropriate. Often we are interested in different stakeholders' perceptions, not just of outcome but also of the way the management change was implemented over time and, in the case of services, patients' experiences during their contact. Controlling for confounding variables prospectively or retrospectively by statistical analysis is not the only way to judge whether other factors apart from the intervention have an influence.

Hierarchies of research design are useful for assessing management research, but we need different hierarchies of design to grade studies of different types of management interventions. The debate about hierarchies of design has begun within the Cochrane Collaboration community, initially in a consideration of research into practice improvement techniques. Meanwhile the research assessment framework such as that outlined above is a useful general method. This approach to management practice is not so different from what some managers do informally: asking colleagues about their experience with or knowledge of management changes or methods, looking up management journals on the subject and ensuring that implementation is monitored and that actions or projects are well defined and structured. The proposed approach is more systematic and disciplined, and congruent with quality management practice.

Will the extra time and effort using management research result in more cost-effective decisions and programme management? In some cases it will not, in part because of the poor quality of management databases and of some of the research: searching often does not reveal helpful evidence. But for high-expenditure or high-risk decisions and projects this approach will be necessary. More importantly, learning the skills, concepts and philosophy of evaluation-informed management is likely to improve a manager's practice and decision making in other situations when there is not time to gather the evidence: a project will be better managed if the manager uses evaluation concepts to clearly define their intervention, the target and how success and failure could be measured.

CONCLUSIONS

Managers and policy makers make interventions in health organisations and systems. Their interventions may not be as immediately life-threatening or life-enhancing as those made by clinicians, but they can cause more long-term suffering and waste more resources than clinicians. Few management technologies, organisational changes or reforms have been evaluated. This chapter has argued that, although there are difficulties, we can and should evaluate such interventions. We need to develop the methods and designs for such research and research-funding bodies need to sponsor reviews of research methods and designs. The chapter has also noted the different research designs which can be used by researchers, or by managers for internal evaluations, and has also described simple methods which managers can use to gain access to and assess such research. It has also proposed that managers can make a greater use of evaluation methods to evaluate their own organisational changes, or collaborate with action researchers to do so.

References

Fowkes, F. and P. Fulton (1991) 'Critical appraisal of published research: introductory guidelines', *British Medical Journal*, 302, 1136–40.
Grol, R. (1996) 'Research and development in quality of care: establishing the research agenda', *Quality in Health Care*, 5, 235–42.
Joss, R. and M. Kogan (1995) *Advancing Quality* (Milton Keynes: Open University Press).
Macdonald, G., C. Veen and K. Tones (1996) 'Evidence for success in health promotion: suggestions for improvement', *Health Education Research*, 11(3), 367–76.
Motwani, J., V.E. Sower and L.W. Brasier (1996) 'Implementing TQM in the health care sector', *Health Care Management Review*, 21(1), 73–82.
Øvretveit, J. (1998a) *Evaluating Health Interventions* (Milton Keynes: Open University Press).
Øvretveit, J. (1998b) *Comparative Health Research* (Oxford: Radcliffe Medical Press).
Øvretveit, J. and A. Aslaksen (1999) *The Quality Journeys of Six Norwegian Hospitals*. Norwegian Medical Association, Oslo.
Øvretveit, J. (1999) *Integrated Quality Development* (Oslo: Norwegian Medical Association).
Pollitt, C. (1996) 'Business approaches to quality improvement: why they are hard for the NHS to swallow', *Quality in Health Care*, 5, 104–10.
Shadish, W., T. Cook and L. Leviton (1991) *Foundations of Programme Evaluation: Theories of practice* (London: Sage).
Shortell, M., J. O'Brien, J. Carman, R. Foster, E. Huges, H. Boerstler and E. O'Connor (1995) 'Assessing the impact of continuous quality improvement/ total quality management: concept versus implementation', *Health Services Research*, 30(2), 377–401.
Smith, G. and C. Cantley (1985) *Assessing Health Care: A study in organisational evaluation* (Milton Keynes: Open University Press).

17 Conclusion
Annabelle L. Mark and Sue Dopson

The purpose of this concluding chapter is to highlight some of the issues and themes emerging from this collection of papers, as well as commenting on those issues that have, as Hamlet said, been 'more honour'd in the breach than the observance'.

The *first* issue, commented on by Dawson, but absent from many of the contributions themselves, is the importance of building on existing theoretical and empirical knowledge. There are many good overview discussions of the choice of theoretical approaches available to researchers, but finding relevant empirical studies is more difficult as such material is scattered across several disciplines with their own associated specialist journals. A recent King's Fund centre project (Mays *et al.*, 1997) seeks to review studies in health care organisation and management that go beyond just descriptive research. A feature of the findings is how few provide rigorous discussions of methodology or research design (as Øvretveit confirms in Chapter 16) and how few go beyond producing descriptive accounts. This is clearly a challenge for those of us working in this area, a challenge which, we would argue, is more likely to be met if we seek to build on what we already know.

The *second* issue which runs through many of the chapters is that of understanding the nature of research (see, for example, the chapters by Fitzgerald *et al.* and Currie) and, more especially, what are the appropriate methodologies. Peck and Secker point out the problems inherent in identifying the audience for research, but argue that these can and should influence the methodologies chosen. This is essential if the research is to influence both practitioner and academic audiences, as Dawson suggests it must, otherwise practitioners may revert to atheoretical satisficing of decisions, often only with the help of management consultants. The satisficing model (Simon, 1956) describes decisions as meeting the most important perceived need based on nearest appropriate options, rather than the ideal solution from all available alternatives for the provision of optimum outcomes. This is true especially where innovation is required to practise to create new forms and methods (Meads, 1997);

for example, in the development of new organisations for primary care. Innovation in such a dynamic social environment requires a variety of social science-based methodologies to evaluate options, yet the NHS has in the main failed to grasp this and continues to structure a hierarchy of relevance in research methodologies with randomised control trials at the top and descriptive studies at the bottom (Baker and Kirk, 1998; NHSE, 1997).

One partial answer to these problems which is not well represented in current research is a triangulated approach. This application of a variety of methodologies to confirm research outcomes is a method well understood in the research community, so why is it not being applied more widely? It may just be a problem of time and limited focus, but there are a number of other reasons which might be explored. In the first instance, triangulation implies collaboration, often across a number of disciplines, to investigate via different research methods whether a proposition is robust. In organisational behaviour, much of the research is qualitative in nature and can encounter credibility problems with some of the more positivist disciplines; indeed collaboration can be quite difficult when the barriers to discourse across disciplines are akin to those of learning a foreign language or, at worst, an undeclared war (Kilduff and Mehra, 1997). A second reason is that, within health care, such research takes place within an organisational culture increasingly predicated on the need for evidence, as explored in the chapter by Fitzgerald *et al*. This also now includes the notion of evidence-based management, which is itself a somewhat problematic concept based on a questioning attitude of mind (Stewart, 1998). The difficulties of agreeing to an adequate interpretation of evidence-based management have led to a situation where it can be used to justify both qualitataive and quantitative approaches, the qualitative researchers suggesting that evidence is often not transferable from one management context to another, while the more positivist researchers tend to concentrate their efforts on looking at only those things which confound this assumption, in the 'if you can't measure it you can't manage it' tradition. The third reason why triangulation is often missing is because of the time-consuming and complex nature of the qualitative research methodologies, which can overwhelm the most intrepid researcher and have high costs in terms of time and financial and human resources.

Furthermore a more positive aspect of this argument about the reasons why triangulation seems lacking is perhaps that, at the heart of many qualitative methods, for example Glaser and Strauss's Grounded Theory (1967), is the notion that the theory or proposition has to be generated *from* the research process rather than proved *by* it. In fact, as Øvretveit suggests in his chapter, many of the most important research outcomes are emergent and therefore not measured from the outset, so alternative methodologies will not even start at the same question, let alone arrive at the same answers; it is not surprising, therefore, that such unpredictability is in part to blame for policy makers' nervousness about evaluation (Baker and Kirk, 1998). What all this means is that, organisational behaviour as a discipline, has few shared templates or standards, but the assumption that it should build only on past perceptions of the purpose of research. In future, it may be that such templates are not meant to exist, but rather that methodologies must be found to capture meaning which continually changes; combining the interaction of feedback from researchers (action research) and feedback from implementation in the management domain (action learning) into actionable knowledge (Argyris, 1997) which enables the knowledge gained to be used to create (but not replicate) intended actions elsewhere.

One other area in which triangulation may be essential in developing an understanding of relevance in choosing methodologies, is in proving whether alternative methodologies themselves are appropriate; the unsuitability of the most culturally acceptable methodology in health care, for example the use of control groups, collapsed when tested against the world of organisational behaviour (Mark, 1993). At a more fundamental level, the dualities often represented as opposing ways of seeing the world of research in social sciences, for example between the subjective and objective approaches (Morgan and Smirnich, 1980), are not just methodological but also represent the philosophical differences between the collective or utilitarian approach and the individualist viewpoints. Objective rationality or the nomothetic research paradigm, upon which much of science is based, assumes that sufficient similarities exist for patterns of transferability to emerge across populations; the subjective or idiographic approach, on the other hand, implies a uniqueness of experience, as encapsulated in the doctor/patient relationship, which some might say confounds the main purpose of a collective enterprise such as the NHS. Yet understanding this unique dichotomy is critical for ensuring the success of many of the key policy initiatives, such as encouraging evidence-based medicine and improving clinical effectiveness.

A *third* general theme is the increasing complexity of the issues those working in this field seek to study. The assumption of a linear process in the progress from one situation to another is refuted by many of the studies set out in this book. Many of the chapters cite complexity, differing interests, complex interdependencies and differing vocabularies that mediate how individuals and groups negotiate their practices and experiences. Such fragmentation and complexity are acknowledged and, indeed, embraced by the post-modern movement (Weik, 1983; Kilduff and Mehra, 1997) and are at the heart of both its strengths and weaknesses (Turner and Pidgeon, 1997). Accepting this approach reaffirms Giddens' (1984) suggestion that 'the uncovering of generalisations is not the be all and end all of social theory', but if it does not do this it must have alternative outcomes which are not yet fully appreciated – not least, the ability to identify the right questions, even if the progress which follows is 'only towards even greater knowledge of our own ignorance' (Kilduff and Mehra, 1997), something those organising health care find somewhat alarming.

Understanding the mechanisms for crossing boundaries between sub-cultures and groups will be essential if improvements are to be made to organisational arrangements, processes and outcomes. The time scales and costs of research in health care often prevent either longitudinal research designs (see Stewart *et al.*, 1988) or multilayered monitoring across the many sub-cultures that can be found in health care, in favour of a focus on particular parts, mirroring the reductionism inherent in the medical model. This situation is not, however, conducive to the integrative policy agenda now being pursued by government.

The focus of much research neglects the relationships with and to others, both in the internal and the external environment, as Ashburner and Birch; Currie; Cropper; Blackler, Kennedy and Reed and Mark suggest in their respective chapters. Such relationships are increasingly important to the effective delivery of health care, and understanding such interdependencies is critical to the development of effective groups and teams. The formation of interdependent groups will not necessarily lead to good teamworking (West *et al.*, 1997), as much of the research in primary care has confirmed. The problems associated with this can be exacerbated by a lack of a managerial structure and organisational strategy, which is owned by all the stakeholders. Such a vacuum, for example as it existed in primary care, is thus filled largely by medical professional interpretations in shaping and designing these organisations and their processes, not always to the advantage of other professionals involved, or the patient. While medical leadership is now at the heart of primary care groups and clinical governance, it may be at a price in terms of stress, as both Morley *et al.* and Alimo-Metcalfe suggest. The active participation of others within this team culture, in leadership roles, has poor organisational precedence, so may prove harder to develop than the political rhetoric implies. Yet, as Ferlie *et al.* (1996) suggest, what distinguishes health care from other types of organisation is that change is dependent on groups and not individuals, so leaders who cannot inspire others to follower will, through ignorance or stress, continue to fail.

Furthermore the focus of much research is on outcomes and the professionals' perceptions of them, rather than patients' or consumers' perspectives. This moves the agenda from organisational behaviour, as defined in our introduction, into organisational analysis, because it concerns what many can dismiss as groups external to the organisation. Consumer perceptions and interactions with the system are explored in different ways in the chapters by Harrison *et al.*, Cropper and Blackler *et al.*, and will in fact determine the future of health care, its roles and relationships, both within and across organisational boundaries, and appreciating the interactive nature of this process is critical to success.

ISSUES FOR THE FUTURE

In completing this review we identify several emerging strands for further thought when considering research in this area. Many of these were also discussed at the original conference identifying what is both absent and present in the current discourse. These ideas were often articulated in simple terms, but in this simplicity lie the germs of an exciting research agenda, and it is for those of us wishing to contribute to elaborate the ideas, in anticipation of the fact that the outcomes will be much like 'shaking a kaleidoscope' (Turner, 1995) in dislodging old patterns, generating new patterns and demonstrating that a variety of alternatives are possible.

- **Collaboration**: interorganisational collaboration, partnership and integration are dominant themes of the new arrangements for the NHS. Understanding how collaborative arrangements form, hold together and perform in whatever policy context exists is critical.

This is particularly important as no one part, or profession, in health care can any longer deliver on its own.
- **Behaviour**: ensuring that research explores what people do, as well as what they say they do. This exploration of behaviour must also incorporate the cognitive and emotional basis for these actions.
- **Consumers**: exploring the organisational role of the user in health services, especially in negotiating the organisational and professional boundaries.
- **Multiple interpretations**: understanding the multiplicity of interpretations and the varying meanings associated with them, including rather more meta analysis where appropriate; together with the impact of new disciplines, for example evolutionary psychology (Nicholson, 1997), which for example seeks to link growing knowledge of our genetic inheritance with behaviour.
- **Context**: the new modes of working must be understood in both the historical and cultural context within health care and the hierarchical structuring of the knowledge which it oversees.
- **Interrelationships**: the need to move away from researching individual policy issues, for example quality, and towards looking at relationships with other significant areas, such as cost-effectiveness.
- **Emergent outcomes**: given the developing complexity of health care, it becomes increasingly likely that there will be a number of unplanned and unanticipated outcomes. These outcomes are important areas for further research.
- **Longitudinal studies**: a reading of the issues set out above indicates the need for longer-term research strategies of at least five to ten years. This is not to say that research with shorter time scales is not of value, but that currently there is an absence of longer-term research that would inform thinking about these very complicated research areas.
- **Conflicting agendas**: the issue of the conflict between managerial and academic time scales will, however, remain in both the commissioning and implementation of OB research. The OB agenda in health is also significantly affected by political as well as managerial and academic timetables in relation to health care, and a balance must be explicitly struck between them. So far, the balance between them has not been considered in anything but a haphazard and arbitrary way.

- **Poverty in pragmatism**: the object of much commissioned research is to find out what does and does not work. However this pragmatic approach often fails to address long-term outcomes which, even in scientific terms, have been shown to reverse the original outcomes (Piachaud and Weddell, 1972). Tracking both across organisations and across time is an essential component of the research agenda in OB.
- **International collaboration**: a final point centres on the opportunities presented in following the example of clinical research in developing international collaborative research agendas, which are currently notable by their absence.

In conclusion developing understanding of the role of organisational behaviour research in health care within the wider research community, whether within clinical research or the wider organisational behaviour research community, will, we believe, enable a productive and exciting contribution to be made to academic, managerial and consumer interests.

References

Argyris, C. (1997) 'Actionability and Design Causality', in Cary L. Cooper and Susan E. Jackson (eds), *Creating Tomorrow's Organisations – a handbook for future research in Organisational Behaviour* (Chichester: John Wiley).

Baker, M. and S. Kirk (1998) *Research and Development for the NHS* (Oxford: Radcliffe Medical Press).

Ferlie, E., L. Ashburner, L. Fitzgerald and A. Pettigrew (1996) *The New Public Management in Action* (Oxford: Oxford University Press).

Giddens, A. (1984) *The Constitution of Society* (Stanford, Calif.: Stanford University Press).

Glaser, B.G. and A.L. Strauss (1967) *The Discovery of Grounded Theory* (Chicago: Aldine).

Kilduff, M. and A. Mehra (1997) 'Postmodernism and Organisational Research', *Academy of Management Review*, 22(2), 253–81.

Mark, A. (1993) 'Researching the Doctor Manager – choosing valid methodologies', *Journal of Management in Medicine*, 5(4), 52–9.

Mays, N., E. Roberts and S. Dopson (1997) 'Systematic reviews of health care organisational and financial interventions', Interim Report for NHS Executive and North Thames R&D Directorate, King's Fund Policy Institute.

Meads, G. (1997) 'The Relational Challenge', *Health Management*, 1(9), 22–3.

Morgan, G. and L. Smirnich (1980) 'The Case for Qualitative Research', *Academy of Management Review*, 5(4), 491–500.

NHSE (1997) 'Personal Medical Services Pilots under the NHS (Primary Care) Act 1997: A guide to local evaluation', Primary Care Division, 97PPO130, Leeds.

Nicholson, N. (1997) 'Evolutionary Psychology and Organisational Behaviour', in Cary L. Cooper and Susan E. Jackson (eds), *Creating Tomorrow's Organisations – a handbook for future research in Organisational Behaviour* (Chichester: John Wiley).

Piachaud, D. and J.M. Weddell (1972) 'The Economics of Treating Varicose Veins', *International Journal of Epidemiology*, 1, 287–94.

Simon, H.A. (1956) 'Rational choice and the structure of the environment', *Psychological Review*, 63, 129–38.

Stewart, R. (1998) 'More art than science?', *Health Services Journal*, 108(5597) 28–9.

Stewart, R., S. Dopson, J. Gabbay, P. Smith and D. Williams (1987–8) 'The Templeton College Series on District General Management', National Health Service Training Authority Publications.

Turner, B. (1995) 'A personal trajectory through organisation studies', *Research in the Sociology of Organisations*, 13, 275–301.

Turner, B. and N. Pidgeon (1997) *Man-made disasters*, 2nd edn (Oxford: Butterworth Heinmann).

Weik, K.E. (1983) 'Contradictions in a community of scholars: the cohesion–accuracy tradeoff', *Review of Higher Education*, 6, 253–67.

West, M.A., S. Garrod and J. Carletta (1997) 'Group decision making and effectiveness: unexplored boundaries', in Cary L. Cooper and Susan E. Jackson (eds), *Creating Tomorrow's Organisations – a handbook for future research in Organisational Behaviour* (Chichester: John Wiley).

Index

Abbott, A. 65
action learning 81–3, 257
active labour management 198
Active for Life campaign 215
activity networks, complex 233–7
  and incompatible priorities
    237–9, 240–1
actor networks 191–2, 193–4,
    211–12
  promotion of physical activity
    213–19; extending the actor
    network 217–19
acute sector
  doctors in management in England
    and the Netherlands 3,
    117–33
  innovations and evidence 194,
    195–201
Alberta, Canada 47, 48–54
Alimo-Metcalfe, B. 140, 144
Allen, P. 49
alliances, strategic 90, 128–9, 172
Alvesson, M. 31
American Medical Association 68–9
analysis
  of data 37–9, 43, 50
  levels of 22
analytical categories 43
Anderson, N. 31
Anthony, P. 224
anti-thrombolytic prophylaxis 195,
    198–9
apothecaries 179
appraisal 141–4
appropriability model 190–1
Argyris, C. 257
Asch, D. 32
Ashburner, L. 64, 65, 70, 73, 120
Aslaksen, A. 249
assessment
assessment centre 144

quality of research 251–2
techniques and leadership 146–7
Atkin, K. 72
Atkinson, P. 164
Atwater, L. 136, 144
audit, clinical 244
Audit Commission 196
autonomous innovation 181, 182–3
autonomy
  crisis of for senior managers
    227–8
  leadership style and stress
    138–40
  professional *see* professional
    autonomy
Avolio, B.J. 136

Baker, M. 257
Barwell, F. 32
Bass, B.M. 136–7, 140–1, 143,
    144, 145, 147
Bate, P. 30
behaviour 260
Berg, P.O. 31
best practice 29–30
'Biesheuvel' model 122–3
Blackler, F. 233–4
Bloor, G. 171, 183
Borrill, C.S. 137
Boss, R.W. 138
boundaries
  alliances across 90, 128–9
  contested 3, 63–76
  managing across 17–18, 21
  research and mechanisms for
    crossing 258
boundary objects 216–17
Bowman, C. 32
Braithwaite, A. 77
British Association of Medical
    Managers (BAMM) 110, 113

Broadbent, J. 182
Brown, A. 68
Buchanan, D. 89
budgeting, hospital 121
Burrell, G. 8, 9
business managers 91, 92, 105, 111

Calgary Regional Health Authority (CRHA) 48, 49–54
Calman, M. 67
cancer services and palliative care team 223, 226, 228–30, 238
career expectations 18–20
Carley, M. 218
Casey, D. 81, 82
categories, analytical 43
Centre for Mental Health Services Development (CMHSD) 36–40
Challis, D. 208, 209
change
　clinical directors and implementing 106–8
　in clinical practice 20–1
　drivers for change and PAMs 171, 174–80
　impact on GPs 77
　managing and coping with 16–18
　mid-career break scheme for leaders 77–87
　need for leadership 21–2
　strategic *see* strategic change
　transformational leadership *see* transformational leadership
Chantler, C. 119
Charlton, J. 137
Chesbrough, H.W. 177, 181, 182, 183
chief executives 82
　and clinical directorates 99, 101, 106, 113
Chief Medical Officer 86
'child B' case 14
Christie, I. 218
Clarke, J. 225
clinical audit 244

clinical directorates 3, 89–116, 119–21, 124–5, 178
　factors in difference 104–9
　managerialist 98–101
　organisational challenge 112–14
　perceived customers 160–1
　power-sharing 101–4
　traditionalist 94–8
clinical directors 91
　organisational challenge 112–14
　recruitment 105
clinical governance 112–13, 131, 178
clinical guidelines (protocols) 176, 190
clinical partnerships 231, 232, 239
clinical practice: managing and changing 20–1
Clinton, W. 14, 174
Cochrane Collaboration 252
Cochrane meta analysis 196
code and retrieve method 50
coding data 38, 43
Cohen, L. 98, 108, 109
coherence 216–17
collaboration 207–21
　as an actor network 211–12
　complex activity networks 233–9
　international collaborative research 261
　and its value 210–11
　policy 170, 172, 174, 223, 239; problematic nature 208–10
　promotion of physical activity 213–19
　research 256, 259
collaborative networks 172, 182
commissioning 69, 74
　see also primary care groups
communication and feedback model 191
communication role for consumer 51–2, 55, 58
competencies
　career development 18–19

competencies – *continued*
  transformational leadership 136–7, 144, 145–6
complex activity networks *see* activity networks
complexity 257–8
  of health systems 11–16, 21
'comrades in adversity' 82
conceptual dimension of research 8–10, 10–11
conference presentations 39, 43
conflicting agendas 260
connection mechanisms 214
Conrad, D.A. 47
consensus building 96–7
consequential value 211
constant comparison method 38–9, 43, 50
constitutive value 211, 212, 219
consumers
  as co-ordinators of care 2–3, 47–62; action roles 51–4; changing roles 58–9; sense making and 54–8
  research and 259, 260
content indicators 22
contested boundaries 3, 63–76
  issues and evidence 69–72
  issues of professional control 64–6
  nursing and medical roles in primary care 66–9
contested evidence 197–9
context 260
  contextual dimension of research 8, 10–11
  diffusion of innovation 192–3, 202–3
  historical 29
  organisational and career development 18–19
continuing professional development 85
control
  integrated services 232–3
  issues of professional 3, 63–76

co-ordination
  clinical directorates 96–7
  consumer's role 2–3, 47–62
coping with change 16–18
Corbin, J. 50
cost-effectiveness 127
court action 14
Crail, M. 39
Cravens, D.W. 172, 181, 182
credibility of research results 199–200
Creswell, J. 50
Cropper, S. 210, 211, 212
cross-boundary management 17–18, 21
cues 55
culture
  aspects of organisational development 30–3
  clinical directorates 107
  marketing and cultural change 159
customers, perceived 160–1

data analysis 37–9, 43, 50
data collection 36–7, 42–3, 50, 157–9
Davies, C. 64
Dawson, P. 171, 182
Dawson, S. 7, 16, 17, 20, 89, 96, 163
de Roo, A. 130
demanagerialisation of doctors 90
demands on time 36, 36–7
Department of Health 77, 137, 172
  The New NHS 63, 84, 131, 170, 173, 208
  nursing roles 68, 69, 74
  Our Healthier Nation 208
deprofessionalisation 90
Deykin, D. 190
dictated narratives 37, 43
diffusion of innovation 4, 189–206
  context 192–3, 202–3
  models 189–94
Dionne, E.J. 14

direction role for consumer 53-4, 57, 58
directorate management teams 91, 92, 104-6
  see also clinical directorates
discretion 138-40
disease prevention 14-15
dissemination model 191
dissemination of research 36, 39-40, 43
distortion of policy 26
district nurses 72
doctors
  control of therapeutic process 232-3
  GPs see general practitioners
  impact of stress 138
  in management in hospitals 3, 117-33
  managing and changing clinical practice 20-1
  medical managers 3, 89-116
  medical organisations in hospitals 124-8
  relationship with managers 128-30
dominance, professional 65-74
domination, institutionalised systems of 9
Dopson, S. 153, 154
downsizing 28

economic change 174-5
Elliott, R. 170
Elston, M.A. 65
emergent strategy see strategic change
enactment of the environment 57
Engestrom, Y. 234
English hospitals 117-33
environment 60
  enacting the 57
epistemological tradition 40-2
epistemology 8
ethnographic research 156-8, 164-5

Esping-Andersen, G. 176
European Union 175
evaluation 5, 243-54, 257
  hospital quality programmes 244-5, 245-9; designs used 248-9
  towards evaluation-informed management 249-52
evidence 4, 189-206
  complex hierarchies of 199-201
  contextually framed diffusion 202-3
  models of diffusion of innovation 189-94
  'scientific evidence' as a social construction 196-9
evidence-based management 256
evidence-based medicine (EBM) 170, 189
  and management 28-30
expansive learning 237
expectations 55
  career 18-20
experience
  career development and 18-19
  of professional practice 200
experimental approach to evaluation 245-8
external managerial roles 17-18

Fairfield, G. 176
fee reductions 121-2
feedback sessions 39
Ferlie, E. 154, 173, 208, 224, 259
  co-operation in NHS 207
  demanagerialisation of doctors 90
  new public management 174, 225-6
Firth-Cozens, J. 138
Fitzgerald, L. 120
Fletcher, C. 141
Florence Hospital 157-62, 163
focus groups 79-80
Fowkes, F. 252
Freeman, J. 214

Freidson, E. 65
French, J. 178
Frost, P.J. 55
Fulton, P. 252
functionalism 65
funds, sources of 8, 10–11

Garfinkel, H. 41
Geertz, C. 41, 158, 171
gender 140–1
General Medical Council (GMC) 180
general practitioners (GPs)
 changing role 63, 66, 77, 84–5
 and diffusion of innovation 201, 202
 mid-career breaks 3, 77–87
 and PAMs 175, 178
 professional control and nurse practitioners 3, 63–76
 relationship with patient 203
generalisability 29–30
Gerteris, M. 59
Gibson, D.V. 190–1
Giddens, A. 258
Gilles, R. 47
Glaser, B.G. 41, 48, 256–7
goal consensus 108
Golembiewski, R.T. 138
government 13–14, 121
 see also policy
Gray, B. 211, 214
Gray, J.A.M. 29
Greenfield, S. 72
Griesemer, J.R. 216
Griffiths Report 65, 154
Grimes, D. 68
Grol, R. 243–4
Grusky, O. 48
Guba, E.G. 49
Gummesson, E. 172
Guy's Hospital, London 91

Hacker, J.S. 14
Hackett, M. 32

Hagedoorn, J. 177, 178
Haines, A. 190
Ham, C. 26, 27
Hammersley, M. 164
Hannan, M.T. 214
Hardy, B. 208, 209
Harrigan, M.L. 58
Harrison, J. 86
Harrison, S. 66–7, 68
Haug, M. 90
health care industry 12, 13
Health and Community Care Act 1990 65, 71
Health of the Nation 213
health practitioners 179–80
health promotion 71
 collaboration in promotion of physical activity 213–19
 resource balance 14–15
Health Service Journal 39
health systems, nature of 11–16, 21
health visitors 72
Hellman, P.S. 138, 139–40
heparin, low molecular weight 195, 198–9
Herk, R. van 117, 127
Hesselink, M. 126–7
hierarchies
 of evidence 199–201
 relationships between health workers 170
 of research design 251–2
Hogan, R. 138
holistic approach to care 67–8
Honigsbaum, F. 67
hospital budgeting 121
hospitals 3, 117–33
 doctors and managers 128–30
 evaluation of quality programmes 244–5, 245–9
 medical organisations in 124–8
 in the 1990s 119–24
Howard, M. 175
Hunter, D. 90, 117, 176, 178
Huxham, C. 210, 217
hybridisation of the NHS 224–6

identity construction 57
ideology 8–10
idiographic approach 257
Iglehart, J.K. 68
ill-health, models of 67
in-depth probing 41, 42–3
incoherence 227–8
incompatible priorities 4–5, 223–42
   collaboration in complex activity networks 233–7
   hybridisation of the NHS 224–6
   implications for research 239–41
   organisational tensions and personal frustrations 226–33
industrial sphere 12, 13
information management 176
innovation 31
   diffusion into practice 4, 189–206
   PAMs 180–4
insecurity 31
'insider' research 41
insurers 121, 123, 129, 131
integrated health care delivery systems 47
integrated services 224, 226, 230–3, 238–9
interactionism 65
interdependence 128–30
interdisciplinary research 74
international collaborative research 261
Internet 43
interpretivism 40–2
interrelationships 260
interventions, evaluation of *see* evaluation
inverted organisations 177
inverted virtual networks 181–4
investment boards 130, 131

Jelinek, M. 191
job control 138–40
job demand 139
Johns Hopkins Hospital 91
Johnson, S. 36, 39

joint venturer clinical directorates 101–4
Jones, M.O. 50
Joss, R. 246
journals 39, 43, 165
   reputation 199–200

Kaluzny, A. 47–8, 172
Keuning, D. 113
Kilduff, M. 256, 258
Kimberly, J.R. 192
King's Fund London Commission 36, 39
Kirk, S. 257
Kitchener, M. 99
Klingensmith, J. 48
knowledge-based networking 177–8
knowledge-driven model 189–90
knowledge management 33
knowledge utilisation model 191
Kogan, M. 246

Latour, B. 212
Law, J. 212, 216, 218
leadership 3–4, 21–2, 135–51, 258–9
   case for more in NHS 137–40
   development 141–4
   mid-career breaks for GP principals 3, 77–87
   organisational development 31–2
   transactional *see* transactional leadership
   transformational *see* transformational leadership
Leadership Questionnaire (LQ) 145, 147
learning, expansive 237
Leasure, R. 49
legislative change 175
legitimacy 214–15
leisure services departments 214, 215, 216
Lessof, L. 69
levels of analysis 22
Lincoln, Y.S. 49

linear models of diffusion   189–92
Llewellyn, S.   108
local experiments   123, 129
Local Government Management Board (LGMB)   145
Lockett, T.   170
Lombardo, M.M.   143
London, M.   141, 142
London Implementation Zone Educational Initiatives (LIZEI) schemes   78
London's Mental Health   39
Longest, B.   48
longitudinal studies   260
low molecular weight heparin   195, 198–9
Lowndes, V.   207
Lunt, N.   72
Lyles, M.   178

'maatschapen'   121, 125, 126
Macdonald, G.   248
Macdonald, K.M.   178
macro capping   121
'manage' role for consumers   53, 57, 58
management   7–24
  approaches to research   7–11
  clinical directorates   3, 89–116
  clinical directors' concept of   108–9
  clinicians in management in hospitals   3, 117–33
  of collaborative partnerships   208–9
  contributions of research to improvement   16–21; career development   18–20; changing clinical practice   20–1; coping with change   16–18
  doctors and managers   128–30
  evaluation of interventions   5, 243–54; towards evaluation-informed management   249–52

evidence-based medicine and   28–30, 256
  special nature of health systems   11–16
  theory and practice   21–3
management of meaning   156
managerialism   73–4, 224–6
managerialist clinical directorates   98–101
mandate of staff executive   127–8
Mark, A.L.   170, 171, 175, 176, 178, 182, 257
market-based reforms   10, 26, 27, 154–5, 224–6
marketing strategy   153–67
  market-driven change and marketing   154–5
  middle managers and   159–61
  setting the agenda   158–9
Marnoch, G.   90, 106, 110, 112, 113
Massey, A.   173
Mays, N.   255
McCall, M.W.   143
McColl, E.   174
McKee, M.   69
McLean, S.   173, 184
Meads, G.   255
meaning, management of   156
Mechanic, D.   175
Medical Act 1983   179
Medical Advisory Committees   124
medical co-ordinator   130
medical directors   120–1
medical managers   *see* clinical directorates
medical model of care   67–8, 170, 232–3
medical organisations in hospitals   124–8
medical practitioners   179–80
medical profession   *see* doctors; general practitioners
medicine: and management   108–9
Mehra, A.   256, 258
methodologies   *see* research
Michael, M.   214, 218–19

microempirical research 163–6
Mid-Career Break Scheme (MCBS) 3, 77–87
  designing 79–80
  evaluation 83–6
  objectives 78
  philosophy 79
  'take a break' and 'time for a change' 81–3
mid-career review seminars 80–1
middle managers 4, 153–67
  feelings about marketing 159–61
  importance in research 162–6
  role in strategic change 154
midwives 200–1
Mills, P. 58, 59, 60
Mintzberg, H. 156, 164, 174, 181
Moberg, D. 58, 59, 60
Moen, J. 126, 130
monitoring role for consumer 52–3, 55–7, 58
Morgan, G. 8, 9, 50, 55, 257
Motwani, J. 244–5
Mulrow, C. 190
multidisciplinary management group 223, 226, 228–30, 238
multidisciplinary research 74
multiple interpretations 260
multi-rater feedback (MRF) 141–4

National Health Service (NHS) 16
  case for more leadership 136, 137–40
  hybridisation 224–6
  importance of middle managers 162–3
  leadership development 144
  New Labour government and 239–40
  reforms 90, 135; hospitals 119–21; market-based 10, 26, 27, 154–5, 224–6; policy, strategy and operational issues 26–8
  transformational leadership 144–7; implications for NHS 146–7
National Specialist Association 125

Netherlands hospitals 117–33
networks 4, 172, 209
  collaborative 172, 182
  of hospitals 231
  inverted virtual 181–4
  in primary care 85
  see also activity networks; actor networks; collaboration
Newman, J. 225
NHS Executive 70
NHS (Primary Care) Act 1997 63
NHS Trusts 112–13
Nicholson, N. 260
nomothetic research paradigm 257
non-managerialist research, relevance of 165–6
Norway 247
Novacare 174, 176, 182
Novanet 176
nurse managers 91, 105–6, 111
nurse practitioners 179–80
  professional control issues in primary care 3, 63–76
nurse substitution 70
nurses 173–4, 178
nursing model of care 67–8, 232–3

objective rationality 257
objectives, clarified 138–40
obstetrics 193, 195, 196–8, 200–1, 202–3
Offerman, L.R. 138, 139–40
Ong, B.N. 90, 118, 128
ongoing property of sense making 55–7
ontology 8
openness 103–4
operational issues 26–8
organisation 7–24
  approaches to research 7–11
  contributions of research to improvement 16–21
  special nature of health systems 11–16
  theory and practice 21–3
organisational analysis (OA) 1

Index

organisational behaviour (OB) 1
organisational context 18–19
organisational development (OD) 1, 2, 25–34
  cultural aspects 30–3
  evidence-based medicine and management 28–30
  policy, strategy and operational issues 26–8
organisational drivers for change 176–8
orthopaedic surgery 195, 198–9
Osborn, R.N. 177, 178
Øvretveit, J. 173, 246, 249, 250
oxytocin 198

palliative care and cancer services team 223, 226, 228–30, 238
Parsons, T. 65
partnership
  clinical partnerships 231, 232, 239
  doctors and managers 128–30
patient choice 202–3
Pearson, M. 70
Peck, E. 36
peer support 85
performance 7–24
performance pressure 138–40
perspective making 236, 237–9
perspective shaping 236, 237–9
perspective taking 236, 237–9
Peters, T. 30
Pettigrew, A. 29, 93, 104, 107, 109, 208
  change process 156, 161; context and 192–3
  co-operative networking 207
Pew Commission Report 171
pharmacists 179
physical activity, promotion of 213–19
Piachaud, D. 261
Pidgeon, N. 258
Piercy, N.F. 172
plausibility 55–7

policy
  collaboration 170, 172, 174, 223, 239; problematic nature of 208–10
  evaluation of interventions 5, 243–54
  primary care 63, 70–1, 84–5, 175
  research and policy making 190
  strategy, operational issues and 26–8
political change 174–5
political sphere 12, 13–15
Pollitt, C. 66–7, 68, 244
Popay, J. 42
post-modern perspectives 4–5, 223–42
Pouvourville, G. de 192
power-sharing clinical directorates 101–4
practical utility of research 8–10
practice
  experience from 200–1
  theory and 21–3
pragmatism 40–2, 260–1
prescribing powers 68, 69
primary care 84–5, 175, 201
  professional control issues 3, 63–76
  see also general practitioners
primary care groups (PCGs) 63, 73, 74, 85, 239
proactive organisational development 32–3
proactivity 103
problem-solving model 189–90
process indicators 22
processual perspective 4, 153–67
professional autonomy 117
  clinical directorates 108–9
  issues between medicine and nursing 65–74
professional control issues 3, 63–76
professional dominance 65–74
professionalisation 178–80
professions 172–3

professions – *continued*
  complexity of health systems 12, 13
  see also doctors; general practitioners; nurses; professions allied to medicine
professions allied to medicine (PAMs) 4, 169–87
  drivers of change 174–80
  future directions 180–4
  UK context 172–4
Professions Supplementary to Medicine Act 1960 173
programmes 231
promotion of physical activity 213–19
protocols (clinical guidelines) 176, 190
public sphere 12, 13
Putnam, R.D. 209

qualifications 18–19
qualitative research 37–8, 49–50
  quality of 40–2
quality
  criteria 44–5
  evaluation of hospital quality programmes 244–5, 245–9; designs used 248–9
  pragmatism and 40–2
question, defining 250
Quinn, J.B. 175, 176, 178, 180, 182
Quinney, D. 70

Ranade, W. 155
randomised controlled trials (RCTs) 199–201, 245–6
Rappolt, S.G. 190
Raven, B. 178
recruitment 104–5
Redman, T. 141, 142
Reed, M. 224
Reilly, P.R. 142
relationships 258–9
  GPs and patients 203

replacement of GPs 86
research 255–62
  approaches to management and organisation research 7–11
  assessing quality of 251–2
  future research agenda and middle managers 162–6
  issues for the future 259–61
  key issues 42–3
  methodologies 255–7; and dissemination 2, 35–45
  obstacles to commissioning/facilitating 35–6, 36–40
  quality and pragmatism 40–2
  relevance of 165–6
  theory and practice 21–3
Research Assessment Exercise 40
resources, best use of 14–15
retrospective focus 55–7
Revans, R. 82
Richards, L. 50
Richards, M.P.M. 193
Richards, T. 50
Rigg, C. 140
Ring, P.S. 209
Ritchie, J. 38, 42
Robinson, J. 89, 90, 106, 156
Rogers, E.M. 190, 192, 195, 204
Roosevelt, T. 14
Rosener, J. 140
Ross, K. 113
Rowley, E. 71
Royal College of Nursing 179
Royal Colleges 124
Royal Commission on Long Term Care for the Elderly 175

Sackett, D. 125
Sailsbury, S.J. 72
Salter, B. 69
Salvage, J. 65, 67, 71
Sargent, L.D. 139
satisficing model 255
Schepers, R. 118
Schofield Report 175, 179, 183
Scholten, G. 121, 122, 123

Schoonhaven, C.B. 191
Schroder, H. 31–2
Schroder, J. 32
scientific evidence, as a social construction 196–9
  see also evidence
scientific sphere 11–12, 13
Scott, T. 119, 131
screening 71
search for relevant research studies 250–1
Secker, J. 41
selection 146–7
Selznick, P. 210
senior managers
  career expectations 18–20
  coping with change 16–18
  development workshop 223, 226, 227–8, 237–8
sense making 48, 54–8
shared skills 171
Sheldon, A. 174
Short Term Home Care Program 48, 49–54
short-termism 110
Shortell, M. 244
Shortell, S. 47–8, 172
Sibbald, B. 175
Simon, H.A. 255
Simpson, J. 119, 131
Singleton, V. 214, 218–19
Skelcher, C. 207
skills, shared 171
Smircich, L. 50, 257
Smith, P. 179
Snape, E. 141, 142
Snee, N. 69
social capital 209
social change 175–6
social factors, and care 203
social property of sense making 55
Solovy, A. 176
Sparrow, J. 140
Spenser, E. 38, 42
Spurgeon, P. 31, 32
stability 216–17

staff convent 125–7
staff executive 126–8
Star, S.L. 216
state 13–14, 121
  see also policy
Stewart, R. 153, 154, 173, 256
strategic alliances 90, 128–9, 172
strategic change 4, 153–67
  agenda-setting for marketing 158–9
  future research agenda 162–6
  market-driven change and marketing 154–5
  middle managers' feelings about marketing 159–61
  middle managers' role 154
  research methodology 156–8
strategy 107
  collaborative 213–19
  policy, operational issues and 26–8
Strauss, A. 41, 48, 50, 256–7
stress
  GPs 3, 77, 84, 86
  leadership style and staff stress 137–40
Strong, P. 89, 90, 106, 156
subjective approach 257
surgery, orthopaedic 195, 198–9, 200
Sutherland, K. 16
systemic innovation 181, 182–3

'Take a break' 81–3
teamworking 32, 258–9
technological change 176
technology 44
Teece, D.J. 177, 181, 182, 183
Terry, D.J. 139
Tettersell, M.J. 72
theory, and practice 21–3
therapy services *see* professions allied to medicine
'thick descriptions' 41–2, 158
Thompson, G. 172
360 degree feedback 141–4

Tierney, K. 48
time
  demands on managers' 36, 36–7
  scales 36, 37–9
'Time for a change' 81–3
total quality management (TQM) 244–5
traditionalist clinical directorates 94–8
transactional leadership 135–6, 137, 147
transactional networks 182
transformational leadership 3–4, 135–51
  dimensions of 136, 145, 146
  gender and 140–1
  in the NHS 144–7
transitions 2–3, 47–62
treatment: balance with prevention 14–15
triangulation 256–7
Trist, E. 213, 214
Trust Management Executives 124
Turner, B. 179, 258, 259

uncertainty 227–8
understanding 40–1
United Kingdom Control Council (UKCC) 70, 180
United States (USA) 14, 174–5, 247
  nurse practitioners 68–9
unplanned/unanticipated outcomes 260

value critical analysis 207–21
  case study of collaboration 213–19

collaboration and its value 210–11
values, shared 107
Van der Ven, A.H. 209
Van Zwanenberg, T. 86
Vangen, S. 217
Vaughan, C. 77
Versluis, J. 126–7
virtual networks, inverted 181–4
virtual organisations 181

Wall, T.D. 139
Walshe, K. 28–9
Waterman, T. 30
Waters, J.A. 156
Watson, T.J. 158, 165
Weddell, J.M. 261
Weed, L. 190
Weick, K.E. 48, 54–8, 60
West, M.A. 31, 258
Wexley, K.N. 143
Whittington, R. 155, 163
Wilkinson, H. 175
Williams, H. 190–1
Williamson, P. 189–90
Wistow, G. 208, 209
Wohlers, A.J. 141, 142
women 175–6
Wood, D.J. 211
work roles, clarified 138–40
Working for Patients 77

Yammarino, F.J. 143, 144
Yin, R. 165